Dagmar
19th. July 1975.

x x x x x

Tyler Whittle was born in 1927, enjoyed a formal education, served for two years in the Royal Marines, and began his writing career in 1955. Since then he has broadcast many radio talks and features and television programmes, written children's books, novels, books of essays on botany and gardening, and biographical novels on Richard III and Queen Victoria. Much of his work has been translated into German, French, Polish and Japanese. He lives in Norfolk, his favourite part of England, and also has a home in the South of Italy. His greatest pleasures are collecting and cultivating plants, reading and bathing.

KU-064-641

By the same author in Pan Books

The Young Victoria
Albert's Victoria

Tyler Whittle

The Plant Hunters

being an examination of collecting with an account of the
careers & the methods of a number of those who
have searched the world for wild plants

PICADOR

Published by Pan Books
in association with William Heinemann

First published 1970 by William Heinemann Ltd
This Picador edition published 1975 by Pan Books Ltd,
Cavaye Place, London sw10 9pg, in association with
William Heinemann Ltd
ISBN 0 330 24487 6
© Michael Tyler-Whittle 1970

Made and printed in Great Britain by
Richard Clay (The Chaucer Press) Ltd, Bungay, Suffolk

Conditions of sale: This book shall not, by way of trade
or otherwise, be lent, re-sold, hired out or otherwise circulated
without the publisher's prior consent in any form of binding or cover
other than that in which it is published and without a similar condition
including this condition being imposed on the subsequent purchaser.
The book is published at a net price, and is supplied subject to the
Publishers Association Standard Conditions of Sale registered
under the Restrictive Trade Practices Act, 1956

Contents

Part Four

The Edge of the World

Appendices

Acknowledgements

The author wishes to acknowledge his indebtedness for the gift or loan of photocopies and for generously giving permission for their reproduction as illustrations to Dr H. R. Fletcher, Regius Keeper of the Royal Botanic Garden, Edinburgh, for No 10. For permitting the reproduction of their copyright material at no charge he is indebted to the Director, the Arnold Arboretum of Harvard University, for No 9; and to the Controller of Her Majesty's Stationery Office and the Director of the Royal Botanic Gardens, Kew, for No 6. For advice and help in finding material for illustrations he is grateful to Miss Mea Allen, whose books *The Tradescants* and *The Hookers of Kew* were stores of information; to Mr R. Desmond, Librarian at the Royal Botanic Gardens, Kew; to Miss Sandra Raphael, formerly Librarian and Archivist to the Linnean Society of London, and to Mr Ian Lyster of the Department of Natural History in the Royal Scottish Museum.

The author is also much obliged to Mr Euan Cox of Glencarse in Perthshire, the plant hunter who accompanied Farrer on his last expedition, both for his fascinating history *Plant Hunting in China*, which was an invaluable source of information, and for kindly giving permission to quote from his book *The Plant Introductions of Reginald Farrer*; to Mr Rolla Rouse of Hasketon in Suffolk for help with particulars about Margary, Forrest and the Chinese province of Yunnan; to Signor and Signora Pinto Antonio for providing a workroom in their garden at Castiglione di Ravello, and to the Reverend Alan Coldwells for his painstaking and careful work in proof-reading and indexing the book.

Formal acknowledgement is also made to the following for licence to reproduce copyright photographs: the Radio Times Hulton Picture Library for Nos 1, 7 and 8; the Controller of Her Majesty's Stationery

Office and the Director, the Royal Botanic Gardens, Kew, for Nos 4 and 5; the Ashmolean, Oxford, for No 2; and the National Portrait Gallery, London, for No 3.

List of Illustrations

To Bridget

Preface

A comprehensive history of botanical exploration has never been attempted, and this potted study makes no claim to merit the description. It is, rather, a brief explanation of why, how and where some plants have been collected, with an account of a few of the better known collectors. The size and unwieldiness of the subject called for such selection and reduction and, as a matter of convenience, it has been looked at as though through the wrong end of a telescope: to make the picture less detailed, much smaller, and yet still clear. It is the attempt of an armchair plant hunter to share his own pleasure in the chase.

T.W.

Introduction

On the Powers of Plants

The Nature of Plant Collectors and Their Motives for Collecting:
the Intrinsic Worth of Plants: Rarity Value: Dahlia and Tulip
Lunacy: Botanical Curiosities: Caste Marks and Economic Plants

IT is well over two hundred years since Carl Linnaeus, Professor of Botany at the Swedish University of Uppsala, took exception to the fact that though deeds of gallantry in his own science were by no means inferior to those of 'kings, heroes and emperors', they were denied an equal share of honour and immortality; and with the sombre seriousness of his race he inquired: 'What labour is more severe, what science more wearisome, than botany?'

This may have been true then. It might still be true of those scientists and enthusiasts who are capable of botanical gallantry. The rest of us -- the large number of fringe botanists – are not required to make such sacrifices. Collecting, sketching, photographing, making herbariums, sowing and planting are continual sources of pleasure; the actual science of botany, a continual source of surprise; the history of our hobby, a continual source of interest.

Being a specialized occupation botanical collecting has had its professionals: men hired by botanic gardens, horticultural societies, governments interested in economic plants, and syndicates of private gardeners; but the amateurs, mostly with a primary occupation like missionaries, consular officials and supercargoes, and others with private means, have also made great contributions to the artificial redistribution of plants.

Amongst them were many fascinating characters: a few Don Quixotes, some sensation-seekers and remittance men; the good, the brave, the knaves, the bad, the mad; those greedy for fame or cash returns; those who were as restless as Noah's raven going to and fro. And over each of them plants have cast certain spells.

Some plants have been collected simply because of their intrinsic worth. Lilies, for example, have an obvious attraction; to the Oriental as table delicacies equalling Chanterelles in flavour; to the Occidental as undisputed beauties associated from antiquity with virginity and sepulchres, with religious cults and royal dynasties. To find and possess Lilies has spurred men to exploits which, depending on one's point of view, could only be described as heroic or utterly foolhardy. One of the greatest of collectors, Ernest Wilson, almost gave his life for the Regal Lily which he introduced to the West from a single desolate valley high in the Chinese province of Szechuan. He and his coolies were returning from the valley through wild hill country where the track was narrow and avalanches not uncommon, when a boulder suddenly dropped from the hillside above them and broke his leg in two places. His description is remarkably detached, but we may imagine his unenviable position: suffering greatly from the fractures, still many days' journey from the nearest medical help and attended only by coolies who would have deserted him if they had thought it worth their while. And no sooner had a splint been improvised from his camera tripod than a fresh hazard confronted him in the shape of a mule train travelling in the opposite direction. It was a large train of fifty mules and the track was too narrow for them to be turned. Nor could the train wait while Wilson's party edged its way, one at a time, past each mule in turn because streams of falling pebbles augured another avalanche.

Wilson was rigid with pain but still conscious. It was imperative for him to get to a doctor without delay and he knew it was impossible to go back. A dangerous solution suggested itself. He told his men to lay him across the narrow track. Then, one by one, the fifty mules stepped over him. Fortunately they were sure-footed beasts, but it must have been a nerve-tearing experience.

Wilson's coolies then carried him on a three days' forced march to the nearest missionary post. Every jig and jog would have hurt him. Carried at an energetic speed over rough tracks and paths in an unsprung travelling chair for seventy-two hours would have been the equivalent of a medieval racking. He lived through it, only to be told by the missionary that gangrene had set in and the leg needed to be amputated. Either heroism – or sheer pig-headedness – made him refuse to accept this advice and, though he lived to say 'I told you so', the infection atrophied and thereafter one leg was shorter than the other and gave him what he sometimes called his 'Lily limp'. Such pain, and a minor deformity, was a high price to pay for 7,000 Lily bulbs, but so large were the powers of this particular plant that Wilson considered it worth while. And Lily-lovers would agree.

The intrinsic worth of plants at once attracts collectors. If they are rare as well they have a second kind of value. The magic of untrodden summits, speed records, possessing the only known specimen of a postage-stamp, exercising the *Droit de Seigneur*, all share the attraction of singularity and rarity, but none can have caused such furore as once-rare plants.

The Dahlia's history is strange. Its origin can be traced to Mexico, where the Aztecs called it *Cocoxochitl*, and in the year of the French Revolution tubers were sent over to a French priest who was chief gardener at the Escorial. He named it after a leading botanist called Dr Dahl but he was less interested in the flowers than the tubers, which he considered might make a tolerable alternative to the potato. The plant, however, became the cause of much envy amongst those who saw it in bloom in the Escorial gardens. Then either a gardener was bribed or a plant bandit was unusually successful. A few tubers found their way to Paris where they were sold to the Jardin des Plantes, but, being put in a hothouse the tubers promptly rotted away. The same happened to a number of specimens sent directly from Mexico to England. They were over-cosseted and went into a decline. From seeds or tubers – it is not known which – Napoleon's Josephine managed to grow Dahlias at Malmaison, and she was exceedingly jealous of her collection. Only her hands planted and weeded and generally cared for the imperial Dahlias, until she had so many plants that it was necessary to put a gardener in charge of them. Inevitably corruption followed. One of her ladies-in-waiting who had asked for a tuber and received a curt refusal was determined to outdo her royal mistress's collection. Her lover was ordered to steal sufficient tubers from Malmaison. Being a melancholy Polish prince he did not attempt to himself, and instead bribed the gardener to carry off a hundred roots. When she heard of it the Empress was outraged. She sacked the gardener, dismissed the lady-in-waiting, exiled the melancholy Pole from court, had all her Dahlias chopped up and dug in, and would never hear the plant named in her presence ever again. It was an expensive gesture because Dahlias were very valuable indeed until, from a third source, that is, from Mexico through Berlin, they were developed and became more common and cheaper. In 1836 a flower-bed of growing plants in good condition was bought for 70,000 francs, and a single tuber was exchanged for a diamond. *Cocoxochitl*, the bright Mexican weed, had come far.

The enthusiasm for rare Tulips at an earlier period was more manic still and a powerful stimulant to plant hunters and gardeners of the day.

The cult had two centres: in the Near East, where the Tulip was the flower of the royal house of Ozman and the Persian symbol of love, and in the trim flats of the Netherlands. Bulbs were sent from Turkey to Vienna in the equivalent of the diplomatic bag, and in time the Netherlands grasped the heart of the market which to this day Holland has never relaxed.

Plant hunters scoured the Levant for wild species. Hybridizers went to extraordinary lengths to produce novelties and were so successful that abruptly – and, curiously, in the middle of the Thirty Years War, the most bloody ever fought – the mania erupted and Tulips became articles of general trade. Speculators met to bargain at the house of the Van der Beurse family in Bruges (from which the Bourse had its name), and they even traded in 'paper tulips', that is, on bulbs that did not exist but were promised by a signed certificate. More than 10 million 'paper tulips' exchanged hands in this way. The real articles were sold in public marts or, in the case of extremely valuable bulbs, by private treaty.

The strangest tales have been told about transactions in the private Tulip trade: how one bulb was paid for with a carriage and pair, and yet another with twelve acres of land; how a merchant in Haarlem paid a fortune for a rare bulb and then, discovering that a near-by cobbler had a Tulip of precisely the same colour, bullied the cobbler into parting with his bulb for 1,500 florins, and stamped on the bulb before his eyes. The cobbler, who liked Tulips, was dismayed; and when, callously, the merchant told him he had been quite ready to pay ten times as much, the poor man realized the full extent of his loss and went and hanged himself.

There is also a singular story which, because it appears in so many different sources, may well be true. It shows the practical perils of trading in such a perishable and expensive item. One Dutch importer was so delighted to hear that a shipment of Tulips had arrived in record time from the Levant that he gave the messenger who brought the news a handsome tip and some red-herrings for his supper. The messenger was a seaman and not over-blessed with brains. On his way out of the warehouse he saw what he took to be an onion lying amongst the costly silks and velvets, and he considered it would serve as a good relish to the red-herrings. It happened to be a 'Semper Augustus', worth about 3,000 guilders, and his supper cost the seaman half a year in jail.

It was with some reason that the English poet Crabbe wrote:

With all his phlegm it broke a Dutchman's heart
At a vast price with one lov'd root to part.

Like the South Sea Bubble, the Tulip Bubble was bound to burst. On 27 April, 1637, a law came into force which obliged Tulip contracts in the Low Countries to be fully met like any other civil contract. Confidence in Tulips waned. One speculator who had been enjoying an annual income of over 60,000 florins did not get out in time. Within four years he had lost palaces, carriages, pleasure grounds and friends. As a beggar in Antwerp he became a living cautionary tale.

Tulip lunacy subsided for a time, to emerge in an exotic form in eighteenth-century Turkey. There the Sultan competed with his Grand Vizier in the splendour of their Tulip *fêtes champêtres*. In one, held at night, the thousands of Tulips were embellished by small coloured lamps carried slowly through the beds on the backs of living tortoises. And, to add a further delight, caged birds sang above the Tulip beds, eunuch gardeners uttered shrill cries to entertain the guests, and a Turkish band played noisily from behind movable screens of shrubs. Really it is surprising that Schubert did not compose a 'Tulip Time'.

By the early nineteenth century American Tulip growers began to flood the London market. So many sound bulbs were introduced that their price dropped to within the reach of working men. They were added to the sextet of Florists' Flowers, and 'Florists' here is not used to describe the shopkeepers and costermongers who simply trade in flowers but immigrant weavers from the Continent and, after them, English artisans who brightened their otherwise drab lives by growing and developing and specializing in certain flowers – and, later, in Gooseberries, Parsnips and Leeks. Their devotion to the Tulip was remarkable. Authentic cases have been recorded of young enthusiasts dying of cold through placing their blankets as a guard against frost over their Tulip beds. By 1849 there was already a National Tulip Show in England and there were three principal sections which must have intrigued as well as mystified non-fanciers:

Byblomens with bottoms white or nearly so; Bizarres
with a yellow ground, broken into a variety of colours;
and Roses, with very perfect cups, cherry, and white,
and rose bottoms.

Beyond the clear claim of some plants to captivate collectors
with beauty or rarity each plant has a claim to the attention of a
botanist. The existence of thousands of withered, flattened collections of pressed flowers testifies to this, as do the herbaria of
botanical and science museums. There is living evidence of a kind
in all recreated habitats out of doors or housed under glass. The
doline in the Cambridge Botanic Garden illustrates how well it can
be done professionally – a keyhole-shaped hole dug in the ground
and planted on all sides with plants only found in the Karst country
of Jugoslavia. There are plenty of amateurs' gardens of a like
character where flower-struck gardeners or science-struck botanists
plant their favourite specimens.

To the botanist a vegetable curiosity is especially worth hunting out. It followed that when the English horticulturalist
John Claudius Loudon found a specimen of Eel-grass in nineteenth-century Venice he was overjoyed and determined to
carry it himself all the way back to England. Eel-grass is
neither striking in beauty nor particularly rare, but at that time
it had not been introduced into England, and as a half-hardy
aquatic with a bizarre method of reproducing itself it had
a scientific interest which captivated Loudon. He carried the
Eel-grass in a tin can over the Alps and on to Paris. There, his
hotel bedroom being stuffy and thinking the plant would enjoy a
little fresh air, he put the can outside on the window ledge. He
roped it down to guard against the awful calamity of it being
blown off the ledge and, confident that all was well, he went to
bed.

All was not well.

Like Southern Italians who casually name the rarest of wild
plants '*insalata*', Loudon's Eel-grass was merely a tasty bit of
greenery to the sparrows of Paris. During the night it disappeared,
presumably going the way of the Dutch seaman's 'onion', and a
mourning Loudon carried his empty can back to England as the
only trophy of what he thought might have been something of a
botanical triumph. In the event he was a little consoled to discover

that the Eel-grass had been introduced to England only a few months before.

Certain plants have another, more disreputable, power to attract collectors; that is, when society makes them into castemarks. It is for this reason that Orchid lovers have always been less numerous than Orchid growers and Orchid wearers. Since the Industrial Revolution the exotic tropical Orchid has been a status symbol, a fact heavily emphasized by Mr Frederick Boyle who wrote in his book *About Orchids* in 1893: 'the plant was expressly designed to comfort the elect of human beings in this age'.

In order to meet the demand for more and more and more, Orchid hunters went to unprecedented lengths, being prepared to travel far through unmapped and dangerous country simply on the hint that a new species or a considerable colony of others might be at the end of their journey; razing thousands of forest trees to the ground in order to collect the epiphytic Orchids which lived like birds in their branches; denuding great areas of Orchid stocks – in some cases actually removing all trace of a species from its original habitat so that it only continued to exist in captivity;* concealing the whereabouts of colonies by using caballistic symbols on their Orchid maps which meant nothing to anyone else, and in some cases leaving forged maps about in the hope they would be copied secretly and send some rival collectors off on a wild-goose-chase.

Side by side with the collecting, hybridizers and breeders worked intensively at home. Propagation was neither easy nor speedy until the recent development of the meristerm system, but nurserymen threw all they had into the job of meeting the demand. At the time of the Boer War one English Orchid grower had a stock of over 10,000 plants. Another sold his entire stock to an American syndicate for £24,000.†

The sale of collections and single specimens was a new excitement in the staid auction rooms of New York, Paris and London. When the long-sought-after Orchid *Dendrobium schroederi* was sold

* In 1847 *Laelia elegans* grew naturally and abundantly on a small island off Brazil. Collectors then said they had never seen such a mass of plants. By 1897 not a single living specimen remained.
† Roughly equivalent in value to £151,500 ($367,273) today.

at Protheroe's in London, the sale rooms were packed because it was a condition of sale that, in order to pacify the native tribe which had first owned the Orchid, the human skull in which it grew had to be bought as well. Such an unusual flower-pot has never been auctioned before or since.

The power to attract hunters of such a remarkable flower as the tropical Orchid was very evident. It stood for so much. To some hunters finding an unknown species was the pinnacle of all their professional ambitions. To the romantic adventurers amongst their number Orchid-hunting was living out a *King Solomon's Mines* existence. To far, far more the Orchid was loot. And quite naturally. Meeting a public demand required the expertise of professionals, and adventurers on the make will speedily attach themselves to any profit-making enterprise.

Economic plants have a special importance. They are, or can be, green treasure to national economics. A first-class example was Kew's successful smuggling of rubber seeds from Brazil, which, ultimately, was highly profitable to the British Government.

The export of such a valuable economic plant as one that produced sap which half-hardened and bounced was prohibited by the Brazilian authorities. There was one way round it. A botanist was sent out from Kew on a 'scientific expedition'. Amongst his collections was a large supply of Rubber seeds. Speed in transport was essential to keep them viable. The Victorians rarely, if ever, did things by halves and he promptly chartered a steamer, the *Amazonas*, in the name of the government of India to carry his 'botanic collections' direct to London from the Amazon. It was an impressive gesture. The Brazilian Customs official noted it. He was charmed, too, to receive a courtesy call from the British consul together with this extravagant charterer of steamers. The two men invited his assistance in the export of a collection of exotic Brazilian plants for Her Brittanic Majesty's Royal Botanic Gardens at Kew. The Queen, they assured him, would be inexpressibly delighted to see the floral beauties of his country. They left with the necessary clearance papers, and within an hour the *Amazonas* was steaming out of Pará harbour with her precious cargo of 70,000 Rubber seeds.

A few, only about 3,000, germinated at Kew to grow into seedlings. Less than 2,000 of these proved robust and healthy enough to be reared and shipped out to Burma and the Malay archipelago, but there they became the parents of all the rubber trees in the Far East, founders of a great industry and living proof that national dishonesty on a large scale can sometimes pay dividends.

These records of successes should not be taken as an indication that plant hunting has invariably met with such rewards. It has not. Nor should it be thought that the pursuit of plants was ever an easy and comfortable undertaking. The case of Ernest Wilson has shown how demanding it could be, and, in the main, plant hunters had to be intrepid, as the following pages will make clear.

PART ONE

Collecting in Pre-Wardian Days
from 1482 B.C.

i

Old Civilizations, the Dark and Middle Ages

Ancient Chinese Gardenmaking: the Putative Age of Certain Plants: an Expedition to the Land of Punt: the Gardens of Persia and Media: Trade in Spices and some Flowering Plants: Aristotle-Theophrastus, Dioscorides and Pliny: Haphazard and Crude Systematic Botanizing in the Dark and Middle Ages: Albertus Magnus

ii

The Renaissance

The Tyranny and Limitations of the Ancient Greeks: Dr von Cube's Expedition of 1485: Bock, Brunfels and Fuchs in Germany: Gesner in Switzerland: Turner in England: Mattioli's Successes in Italy: Busbecq at the Court of Suleiman the Magnificent: Dodoens, l'Obel and l'Ecluse from the Netherlands

iii

English Collectors of the Orangery Era

Royal Gardeners: Herbalists and Apothecaries: some Distinguished Collectors of Wild Plants: the Infectious Passion for the Orange and *Choice Greens*: John Evelyn's Influence: Patrons and Plant-Hunting Parsons: a Bloodthirsty Botanist

Old Civilizations: the Dark and Middle Ages

PRE-HISTORY often wears a mad, *Through the Looking-glass* aspect. Europe, for instance, in many features was the exact opposite of what it is today. The site of Rome and all the great central plain of Italy lay at the bottom of a sea. Where England is there was an ice-cold ocean, its surface so thickly packed with floating islands that they jostled one another. The Harz Mountains were islands, too; and what are now the islands of the Greek archipelago were the hills and mountain peaks of a lush plain grazed by rhinoceros and hairy mastadon.

What these beasts grazed on is only partially known. There is geological information about certain prehistoric plants; the Rose fossils, for example, found in Oregon and Colorado which are at least thirty-two million years old. But it is slender. Moreover, it lies beyond the province of this inquiry for no one can know whether or not there was plant hunting in palaeolithic times. Maybe, one day, fresh cave drawings will be discovered which record that amongst the usual scenes of primitive husbandry, concupiscence and the chase man had the wisdom to search out economic plants. Until then we may conjecture that he lacked that wisdom.

Even within history some of the earliest known facts are fairly bizarre; few more so than that five thousand years ago the Chinese had fine gardens with pagodas, aviaries, torrents, mist-covered mountains, still lagoons, formalized bridges, and artificial islands of 'youth eternal' and 'never-ending content'. This was long before any sort of civilization had begun to seep into Europe through the keyhole of Crete, and the Chinese Emperor Chin Ming was cultivating Roses and other flowers when the Greeks had not even reached Greece.

Such facts, considered however calmly, are still astounding. Equally so is the age – or putative age – of certain living plants. That some of the Swamp Cypresses seen today in the United States were actually mature when Moses was inundating Egypt with locusts, blood, boils and frogs, gives them a weird aura. It is shared by the Great Redwoods of California which were already hoary with age when Caesar was assassinated on the Ides of March.

Equally difficult to appreciate is that the first recorded plant-hunting expedition took place when Athens, Rome, Paris and London were non-existent. It was mounted by Hatshepsut, Queen of Egypt* who wanted frankincense, the gum of a tree now called Boswellia after Dr Johnson's randy biographer. It grew on the shores of East Africa, the land of Punt, and five ships sailed south to return with quantities of seeds and plants, many Boswellias and other living trees. Thirty-one of the trees were successfully established in the Queen's Temple Gardens at Karnak where an official record was carved on the walls to mark the success of the expedition.

It was from the baked land of Sumer at the mouths of the Tigris and Euphrates, and from Egypt on the Nile, that Western civilization drew its origins. It was from these sources as well that Westerners drew their love of flowering plants. Persia and Media in particular had a profound influence on the gardenmakers of the West. The Persian love of trees, their parks (*paradeios*) where fountains played and carefully-bred deer cropped nurtured herbage, their enclosed oasis gardens, and their vast Rose gardens fathered imitations which can be seen today in places as far apart as Hull and Honolulu.

As man grew in sophistication his need to douse bad smells and flavours was met by merchants who traded in spices over the ancient caravan routes. And certain plants, aromatic or fragrant, or simply decorative were also included in the freight.

Violets were required because they were the basis of Persian sherbet and the main constituent of a sybaritic Roman dish described by Apicius as 'violets fried with slices of lemon and

* A thousand years before, in 2500 B.C., a pharoah named Sankhkere had sent a fleet down the Red Sea to search for supplies of prepared spices, but it could not properly be described as a plant-hunting expedition.

orange'. They were also used extensively as chaplets all through Greece and Magna Graecia.*

Rose roots were also traded between Rome and the Rose nurseries of Greece at Cyrene and the far-distant nurseries in Persia and Afghanistan, because besides being used as decoration – for carpets and pillow covers – and as cushion stuffing, they were drunk as Rose wine, bathed in as Rose water, and eaten as Rose jelly, Rose honey, candied Rose petals, and a Roman Rose pudding which included, amongst other ingredients, calves' brains, pepper, eight eggs, and oil.

Certainly the Paeony travelled from East to West. Pliny the Elder called it the oldest of all cultivated flowers and it had been growing in China for centuries before his time. Even today it is still so much an essential part of China that it appears on their postage-stamps.

Who first collected the different species of these genera is not known, but it may be taken as a certainty that they were not professionals. In large degree the first plant collectors were amateurs.

In time there was a need for some sort of catalogue of plants. China had her ancient, though far from complete Materia Medica, but no sort of list had been drawn up in the West.

Then, in the fourth century B.C., Aristotle began his observations of nature. His two botanical works were lost – and what were called 'Aristotle's Works' were really a summary of the originals written by one of his pupils called Theophrastus. Moreover, the 450-page summary passed through other hands and languages; being translated from Greek into Syrian, from that into Arabic, thence into Latin, and finally (as late as 1350) being translated back into Greek as *Aristotle's Works*.

Four hundred years after Theophrastus, Dioscorides, physician to Antony and Cleopatra, wrote a descriptive list of six hundred medicinal plants. It was to have a massive influence on botany for over a thousand years, as did the writings of Pliny the Elder who was so passionate a naturalist that scientific curiosity took him to investigate and be immolated to a cinder by Vesuvius in eruption.

Aristotle's pupil, Alexander the Great, demonstrated that foreign campaigns could be useful to the plantsman. From his

*Pindar's descriptive adjective for Athens was 'the violet-crowned' and chaplets of Violets were always on sale in the Athenian markets even in the depths of winter.

Eastern wars Europe gained the Double Yellow Persian Rose and the Lombardy Poplar. And the soldiers of all nationalities, as well as common sailors, merchants, and other travellers carried plants from place to place either because they were beautiful or good to eat or useful as medicines, or in the case of imperial legionaries on a tour of duty abroad, simply because they were reminders of home. In this way the Madonna Lily spread from Roman camp to Roman camp right through Europe – as a wound herb for the surgeons' use and a nostalgic souvenir of home to the cohorts under their care. In this way, too, amongst many other esculents, the Cucumber and Radish travelled north to feed and please the Roman administrators who ruled Britain for over 500 years. Within degree, all these and many more collectors could be described as amateur plant hunters.

A general interest in plants, like scientific curiosity and many other things, fell into desuetude when the Goths initiated the Dark Ages. By then cultivating herbs had become a medical affair, the job of monks and apothecaries' assistants, and herbals such as *Bald's Leech Book* were written with more romance than fact between their covers. But the English paid small attention to flowering plants or any sort of gardening from the time Rome withdrew her legions until the Norman Conquest. Then, true to the long-established tradition which allies horticulture and the clergy, priests from Normandy reintroduced gardens and the art of cultivating flowering plants to their fellow priests in England.

The great Charlemagne was a noted royal plantsman. Crusaders carried home herbs from the Holy Land to sweeten their dirty castles in Western Europe. Pilgrims, scalds, troubadours, pedlars and other adventurous travellers collected and added to the store of known plants. But more and more it became the province of the clergy to maintain supplies of plants, to study them systematically and record its observations. One of the most remarkable was Albertus Magnus, who lived from 1193 to 1280. He stood out amongst medieval botanists as a moon against night-lights and deserves examination as a worthy representative of pre-renaissance plant hunters.

Albertus's long life, if nothing else, marked him as unusual. To live eighty-seven years in a period when very few people reached the age of fifty was in itself an accomplishment. And he was one of those rare beings who juggle with their waking time, managing three men's work in a day, and showing proficiency in many arts and skills. Added to all this he was a most uncomplicated, humble man and he had an ear for anyone.

Precise information was a rare commodity in his day. News travelled slowly and was embroidered on the way. This accounts, perhaps, for the fact that minor contradictions and inconsistencies appear in the recorded events of Albertus's career. But they are small and do not amount to much. On the bare bones of his long and colourful life every authority would agree.

He was very tiny but, though like that other famous dwarf, Toulouse-Lautrec, he had a title as Count of Bollstädt, he was merely little and not at all deformed. He went to Padua where the university authorities were sufficiently enlightened to make the first botanic garden in the world and he enjoyed being a student. Formal disputations and informal talk helped to satisfy his natural inquisitiveness and his great thirst for knowledge. The average student applied himself to theology and philosophy and left it at that. Not so Albertus. He mastered the set curriculum and then his inquiring mind took him to investigate other matters – trees, plants, stones, beasts, fishes, the stars and the elements. He believed they deserved close scrutiny as parts of God's creation. From him they got it.

In time he became a Dominican friar, teaching in the schools of Germany and France, and inspiring Thomas Aquinas at the University of Paris. Fortunately for natural history his teaching career was brought to a temporary halt when he was elected a Provincial of his Order and he had the responsibility of visiting and inspecting each monastery in the province. By a strict application of his vow of poverty, he walked everywhere on foot and, as a result, he botanized over large parts of Northern Europe, his eyes so open for natural history specimens that he became an expert in zoology and mineralogy as well as botany. It was something of a feat of endurance and, in view of the violence of the age, it was a miracle that he managed to survive it at all. To wander, collecting, over tracks and roads infested with cut-throats and live to write seven long volumes on natural history showed that he was particularly blessed.

Albertus's name was soon well known. His wisdom and intellectual brilliance were in demand by his Order, by the universities and by contending parties who trusted him as a mediator. Eventually he was offered an attractive post of lecturing in Rome, but no sooner had he established himself than the Pope decided he was just the man to send on a delicate mission to Poland. The Poles were practitioners of euthanasia. Crippled children and the unwanted old were snuffed out and often eaten. Albertus was charged to make a firm stand against this piece of Baltic barbarism. This he did with great success, and, botanizing his way back to Rome, asked to be allowed to retire to his cell and there contemplate the splendours of God's creation and the certainty of his own speedy end. After all, he was sixty-seven.

The Pope was clearly a man of vision. Not only did he refuse Albertus's request, but he made him, willy-nilly, Bishop of Regensburg.

It was on this famous occasion that the smallness of the new bishop was made evident – for the Holy Father was considerately telling him to stand and not weary his old knees when he realized that Albertus was already on his feet and standing upright. It speaks much for both of them that, while cardinals, chaplains and papal officials froze with embarrassment and pretended not to have heard, the Pope and the bishop broke the silence simultaneously with loud laughter.

Albertus was too humble a man to realize what he had achieved with his writings on husbandry, and over generous in his acknowledgement to Aristotle (that is, to the Latinized version of Theophrastus). He was limited to the extent that he separated all plant life into only two categories,* and he was optimistic to attempt a European flora which, because of the large variations in plant habitats, has not been done to this day; but he was the first to hunt plants systematically over a large area, and the first to publish accurate descriptions of foliage and the importance of a worked tilth and feeds of manure to successful agriculture. He was precocious, as well, in aesthetics, showing an appreciation of water, shorn lawns, and the colour and fragrance of flowers in the green pleasure grounds he described which was far in advance of his time.

If Aristotle may be counted the founder of botany, Albertus Magnus deserves to be remembered as the first systematic plant hunter known to history.

* Theophrastus, Dioscorides and Pliny had separated plant life into three categories. The botanists of the Renaissance varied between twenty-five categories (Dodoens) and no category at all (Brunfels).

The Renaissance

IT is extraordinarily difficult in an industrial age and from this distance of time to estimate the extent of influence which Dioscorides, Pliny and Aristotle – Theophrastus shared between them until the sixteenth century. That is, of course, in the Old World. Across the Atlantic, happily ignorant of the old learning and therefore untouched by the new, Mexican gardeners were cultivating a large number of herbs whose curative and prophylactic properties had been known for centuries by the Aztecs. But in Europe both gardenmakers and botanists were in thrall to the ancients, and it required large, open minds to break free.

One of the first to understand the limitations of the old Greek botanists was a rich doctor and amateur botanist from Frankfurt called von Cube. He had an idea – it was no more – that the flora of Northern Europe differed from that of Southern Europe and beyond. He chose to experiment. And in 1485 he set out on a plant-hunting expedition accompanied by a botanical painter 'of good sense, skilful and cunning'. Their itinerary was ambitious – Italy, the Balkans, Crete, Rhodes, Cyprus, the Holy Land, Egypt. And their plant hunting was successful. Dr von Cube proved that the Greeks of two thousand years before had not written the last word on plant distribution.

His work was consolidated by three fellow-countrymen: Heironymus Bock, Otto Brunfels, and Leonhard Fuchs. They were all from humble homes, all young and anti-authoritarian when Luther nailed up his Theses, and therefore all were sympathetic to protestantism. Their work as plant hunters and classifiers was so important that it appeared to alter the science of botany altogether.

To begin with they loyally tried to squash all the plants they collected into the old Greeks' straitjacket. But reason as well as their age made them rebels. Their new discoveries *were* new. Listing them in herbals, with notes on their description and use, became a lifetime's work.

They had their counterparts in other countries. One of the best botanists and most unlucky of them all was the Swiss Konrad Gesner. His father, a furrier, was the victim of the bitter religious wars between the Swiss cantons. Gesner himself, though an exceptionally industrious botanical collector and physician, was obliged to take on hackwork, and he never knew what it was to be comfortably off or see his work as a botanical collector in any way commended. Towards the end of his life he was given a university chair of medicine, but it was simply an extra burden bringing him next to no salary. He could barely afford to travel to Vienna when summoned by the Emperor Ferdinand. His one extravagance was to have woodcuts made of fifteen hundred of his own plant drawings. These with a huge portfolio of notes, his own observations of plants and his analysis of them into categories, were to be the foundation of a flora 'untouched by Greekery', and, therefore, entirely unique. Misfortune prevented it. Before any of the work was published he contracted the plague and died. His portfolio, the woodcuts and the drawings were sold for a beggarly 175 guilders, the sum being used to pay the engravers, and the purchaser, a botanist named Camerarius, who later was to prove that plants had sexes, and who used the drawings to illustrate his own writings without any acknowledgement to Gesner. It was not until 1751 that a private enthusiast called Trew ferreted out two-thirds of Gesner's woodcuts, and published them under the title of *Gesneri opera botanica.*

The first most considerable botanist in England was William Turner, a clergyman so bigoted in protestantism that he was deprived of the Deanery of Wells and obliged to live abroad in less congenial Cologne. There he practised medicine, having taken a medical degree in Italy, and waited patiently for eight years until the two protestant burners, Henry VIII and his daughter Bloody Mary, were safely dead. After two smaller publications his three-volume *New Herbal* listed about 240 English plants with their localities, and was the first botanical work to be printed in English. This has earned it a position of importance in the Natural Science Museum in London. Turner was essentially a 'local' botanist, helped by his family, most of all by his son Peter who became an M.P., both of them being painstaking, neat workers and thorough hunters over a small area. Only one introduction is credited to Turner, that vastly useful exotic economic plant called Lucerne or Alfalfa. He died in London, still an absentee Dean of Wells, though on this occasion by choice, in the year 1568.

An almost exact contemporary who met with outstanding worldly fortune was the Italian Pier Mattioli. He had the sort of sunny disposition which conceals roguery and appears to invite success. His commentary in Italian on Dioscorides sold 30,000 copies without any trouble at all. A Latin version could hardly be expected to have the same appeal, so Mattioli wisely decided to illustrate it with 500 woodcuts by the popular artist Weidnitz. As a result, reprints of both versions were twice ordered. The book was even published in the Near East, in Syria and Persia, and still copies were so highly valued that they were given in dowries and bequeathed in wills.* Understandably perhaps Mattioli's head was swollen. Less understandably, vanity made him ungenerous, even vindictive to the few who had the scholarship and the courage to point out his mistakes. At his instigation one critic lost his teaching post at Padua University. Another critic found himself being hounded by the Inquisition because Mattioli spitefully drew attention to the fact that he was an orthodox Jew. Because of these flaws in his character it is difficult to admire Mattioli as a man, and because he stuck so closely to the profitable last of Dioscorides, it is difficult to gauge his importance as an innovator. But the botanists of his day held him in high regard and, having wide territories of plant life open to him in Italy which were denied to others, his contribution to the botanical new learning and the collecting of dried herbarium specimens and living plants would have appeared particularly impressive. Certainly he was respected by his contemporaries; being honoured with titles and addresses, compensated by a public subscription when his house was burnt to the ground (and so handsomely that he found himself well in pocket after the rebuilding), and appointed personal physician to the Holy Roman Emperor, Maximilian II. Mattioli 'arrived' in a sense that no other Renaissance botanical collector was ever able to match – though one of his friends and an excellent plant hunter on his own account, came close to enjoying the same fame.

* One copy was bequeathed to Henrietta Maria, Queen of England.

Ghislain de Busbecq was a Fleming and Imperial Ambassador at the court of Suleiman the Magnificent at Constantinople. It gave him great opportunities to hunt out the plants of Asia Minor which, in the sixteenth century, was the chief supplier of Europe's bulbs and stocks of flowering plants. In one year alone Busbecq introduced the Horse Chestnut, Lilac, Tulip, Mock Orange, the Syrian Rose Mallow, *Hibiscus syriacus*, and *Hyacinthus orientalis*; and he was constantly supplying his lucky friend Mattioli with fresh plants and information.

Of the remaining botanical collectors stimulated by the Renaissance three were outstanding. By a coincidence all were Netherlanders, all were close friends, all came from comfortable circumstances, and all had a profound influence on European gardens. Their names were Latinized as Dodonaeus, Lobelius, and Clusius, but they knew each other as Rembert Dodoens, Matthias de l'Obel, and Charles de l'Ecluse.

Dodoens is best remembered as the gull whose work was (badly) translated into English by John Gerard, and hashed-up, with barely a pretence at disguising the fact, as by far the larger part of the famous 'Gerard's *Herball*'.*

L'Obel is mostly remembered for the flower which has his name. It is forgotten that he earned renown in England, where he lived for almost thirty years, marrying an Englishwoman and being appointed King's Botanist to James I. He was the first to write a detailed description of all the Roses cultivated in Europe in Shakespeare's time; that is, of ten species.† He also wrote a history of cereals, discovered the Heath Rush both in England and Wales, and was the printer of an exuberant pamphlet entitled 'The Fullest directions for the Brewing of the most agreeable of all German and English cervisia, or Beer'. He managed to quarrel with Gerard. He recognized the Englishman's prowess as a plant hunter and gardener, for Gerard had noted seventy new plants and cultivated more than a thousand flowering plants, but l'Obel also saw that 90 per cent of 'Gerard's *Herball*' was the work of his old friend Dodoens,

---

* Equally, very few of the 1800 illustrations were Gerard's – having been lifted wholesale from a German flora by Jacob Theodore of Bergzabern.
† It is interesting to note that Pliny the Elder had described twelve kinds of Rose, but Albertus Magnus only four, and that today it is generally accepted that 120 different species grow in the northern hemisphere.

and that even parts of his own notes had been included in a faulty translation from the Latin. It is barely surprising, though apparently it surprised Gerard, that after the publication of the *Herball* l'Obel became 'a hard man' towards the plagiarist.

Charles de l'Ecluse was the most enthusiastic plant hunter of the three friends, but this did not put his interests in blinkers. On the contrary. Probably more than any other contemporary botanist he was the personification of the Renaissance man: studying law at Louvain, and philosophy at Wittenberg to sit under Melanchthon and Luther, and then in turn at Frankfurt, Strasbourg, Lyons, and Montpellier; being fluent in eight languages; at ease with the distinguished and the great, with Rabelais and with cardinals although he was a passionate protestant; noted as an historian, map-maker, mineralogist, zoologist, philosopher, numismatist, and most of all as a botanical collector. He belonged to the Flemish quality, his father being Seigneur de Watènes of Arras, and this gave him the necessary *cachet* to accompany the Fugger brothers, Johann and Markus, when they travelled through Europe on a grand tour before the days of the Grand Tour. In Spain alone l'Ecluse collected 200 new plants which he wrote up and published under the title of *Rare Spanish Plants*. He travelled, as well, in Turkey, collecting and introducing to Europe the Fritillary, a selection of Daffodils, Orrice Root, *Iris florentina*, and reintroducing the Ranunculus and Hyacinth; and later in his career – often through curious roundabout means – he was responsible for the introduction and general distribution of such noted plants as the Tuberose (which he received from India though it is a native of Mexico), the Scarlet Runner Bean from Brazil via Portugal, and the Potato from Spain (whence it had travelled from Peru). But his plant hunting, as a younger man, in Central Europe was his most remarkable contribution to Renaissance botany. And these were not merely the achievements of a dilettante with infinite resources and plenty of time on his hands. In his up-and-down life l'Ecluse knew poverty and riches in turn, and he wore both with a gentle dignity which won him a large circle of admiring friends. He was also caught up in the Low Countries' revolt against Alba. One of his uncles was murdered in the punitive massacre which followed the revolt. His own father lost everything and fled to Antwerp. There l'Ecluse took care of him until, mercifully, he died in 1573. By this time he was in such straitened circumstances that he was grateful to be offered the post of supervisor of the Imperial Gardens in Vienna. From this 'low' in his career, he rose to be knighted by the Emperor, given a professorial chair at Leyden, and to become an established European figure. His most famous work, *Rariorum Plantarum Historia*, was published when he was seventy-five. To his friends there he said that no matter how cruel the vicissitudes of life could be, botany and gardening were everlasting solaces.

3

English Collectors of the Orangery Era

In the University Botanic Gardens at Cambridge there is a serpentine border of mixed perennials, biennials and annuals, not planted according to family or size or colour or flowering season or any of the usual reasons, but, ranged in order from north to south, according to the date of their introduction to England. For this reason it has more of an academic interest than anything else; though it is full of surprises for those who associate, say, Lamb's Ears, Coral Bells and Stocks with Merry Old Medieval England, and discover they have only been in England for two centuries at the most; and it is equally surprising to those who because of the exotic appearance of Bear's Breech and the Winter Flowering Cherry have considered them recent introductions, and find instead that both plants were hoary 'natives' long before the Hundred Years War. The bed makes no claim to be complete. It is merely illustrative and so it does not show the plant-introduction explosions of the seventeenth and nineteenth centuries. The first coincided with the interest shown by a cultivated few in massive orangeries of masonry; the second with a general and less cultivated interest in the huge glasshouse ranges of Victoria's time.

The seventeenth century in England saw her Renaissance in botany. Every art and science from sonnet-writing to plumbing was revitalized by the New Learning, but at a different pace from place to place. When, for example, on the Continent, gardens were altering their character to be no longer repositories of 'rare plants for the contemplation and admiration of nature', these very plants were falling under the notice of English gardenmakers for the first time. French gardeners were interesting themselves in logical progression from vista to vista, in form, shape and texture, in dim grottoes beneath raised terraces, in geometric designs, in parterres of water, coloured pebbles and brick dust, in *compartiments de broderie* so exquisite that, as de Serre, the French farming writer, noted, 'There really is no need to go to Italy!' The French kings had their 'royal embroiderers' to superintend gardens, and the business of flowering plants was really in the province of the *peintre ordinaire de sa majesté pour la miniature* who, more like a floral taxidermist than a plantsman, preserved specimens with paint.

Simultaneously the English were interesting themselves in simples, oranges, and a few 'choice and tender greens'. Slow to mature as ever, the English did not catch the excitement of botanical collecting until long after everyone else. Having caught it, they applied themselves so vigorously to botany and cultivation that within a short space of time they had reached a position of excellence which they have yet to lose. And this they did through one of the most turbulent centuries in their political, social and economic history, when Charles Stuart lost his head and his son James II lost his throne, when, while 'noble reason' stirred men's minds, witches were hunted out and pricked and legally tortured, when religion was often a matter of politics, and vice versa: a century which stretched from *The Tempest* through *Paradise Lost* to *Absalom and Achitophel*, and the pendulum swung to and fro between riotous licence and puritan severity.

It is astounding to think of the English plant hunters working calmly and efficiently in such an age of lace, logic, blood and bigotry.

The Tradescants, father and son, set the pace. There was a third as well, grandson of the first, and all confusingly called John, but Old John and Young John were the great collectors and gardeners.

Old John left his native Suffolk to garden for Queen Elizabeth's Secretary Cecil, Lord Salisbury; and, having laid out the grounds and supervised the main plantings at Hatfield, he went off to Europe in 1611 to search out 'curious greens'. Naturally he went to the Low Countries for stocks of bulbs and new varieties, and went on to Paris for further supplies.* But this was simply his job. Private plant hunting out in the field was his first love, a point he proved later by signing on as a gentleman volunteer against the

* Included in his swag was 'an exceedingly great cherye' – now listed as Tradescant's Heart or Oxheart.

corsairs, when he ran the serious risk of being taken into slavery, so that he would have the opportunity to hunt plants in North Africa. The 'Algiers Apricocke' is one trophy from that hare-brained expedition. In fact, Old John hunted in places as far apart as Archangel, where he saw many Russian flowers including '*Helebros albus*, enough to load a ship', and South Devon, where on a widow-woman's garden rubbish dump he discovered that odd fruit the Plymouth Strawberry, *Fragaria muricata*. It had been rejected as abnormal and, though its descendants are still some-times seen today, the variety does not appear in commercial fruit lists because the flower petals and sepals combine in the shape of a choirboy's ruff and the bright red fruits have miniature leaves protruding from them in the place of pips.

Old John was a respected friend of Jean Robin,* curator of the Jardin des Plantes, after whom the Robinia was named, and they exchanged botanical gossip as well as seeds and vegetative parts. Charles I chose him as his gardener and after him, Young John, who, well-trained in his father's ways, was at one time a successful hunter in the new colony of Virginia. His introductions from the New World were highly valued and numbered amongst them Lupins, Cornflowers, Michaelmas Daisies, Bergamot, Virginia Creeper, Swamp Cypress, Red Maple, Canadian Columbine, and the Spiderwort which botanically bears his name, *Tradescantia*.

Young John remained King's Gardener, working on his cata-logue of all the Tradescant collections and quietly cultivating his private garden of marvels at Lambeth, right through the Civil War. He was so detached that he barely seemed to notice the hurly-burly that went on outside his garden walls, not even the beheading of his master which made him, willy-nilly, for a time Parliament's or the Commonwealth Gardener.

That it has been worth a publisher's while to reissue some of the old herbals in recent years, and that out-of-date, unsystematic pharmaceutical treatises are still taken over-seriously by zealous herborists, suggests that Anglo-Saxons, more than any other race, are prone to herb crankiness. Its degree would have to be multiplied many times to estimate how much value the seventeenth-century English attached to wound herbs, balms, fever alleviants, anodynes,

---

* Robin also sent over curiosities for Tradescant's Ark, a collection which later was to be filched by a contemporary and scientist named Ashmole. He, in turn, bequeathed the loot to Oxford.

astringent and antiseptic herbs. A limited number of vegetables and fruits were grown at the time, but to the English 'plants' meant 'herbs', and it is noteworthy that between three and four hundred medicinal and culinary herbs were listed as the contents of a contemporary English garden, with barely any flowering plants admitted as mere decoration.

Inevitably, then, as the religious houses in England had been shut and looted, and as gunpowder had made obsolete the defended castle and moated grange, both of which had contained herb gardens, it fell more and more to apothecaries to accept herbalism as their special province and they became the practitioners of botanical knowledge.

It was the botanist's interest to hunt plants and classify them; the gardener's to hunt and cultivate them; the apothecary's to hunt and utilize them.

Doyen of the London apothecaries was John Parkinson of Ludgate Hill, whose book with the punning title *Paridisi in sole Paradisus terrestris** constituted the first dictionary of gardening. In the main it listed plants in cultivation, but his description of many wild plants showed that he was also a diligent hunter out in the field. He never travelled very far from London because it was unnecessary. An amazing number of genera with their species were then growing and thriving within sight of the Tower.

Parkinson's careful work over many years was written up in his second book *Theatre of Plants*, and announced for publication in 1640. Because, clearly, it would be a brand new and important herbal the announcement alarmed the printers of 'Gerard's Herball'. They knew they could only hold their public by issuing an up-to-date and revised version of the original, and hired another apothecary, Thomas Johnson, to work over the text and make the necessary amendments. The plethora of mistakes made it a formidable undertaking, and it was made the more difficult for Johnson by the imposed time-limit of twelve months. Nevertheless he managed it and the revised *Herball* was issued before Parkinson's new book came out. Doubtless their printers were at loggerheads but no rancour existed between the two apothecaries. This was an unusual feat of good-naturedness, but, then, both men were widely respected for their affability to everyone.

Johnson was one of the most attractive personalities in the whole

* The Earthly Park of Park-in-Sun.

gallery of seventeenth-century plant hunters. He was a Yorkshire-man, born in 1600 in the town of Selby, and like many before and since he had the initiative to travel up to London where a wider circle of acquaintances and sharper wits made for a more fulfilled life. He became a member of the Society of Apothecaries. His shop in Snow Hill was to become famous when he exhibited the very first bunch of Bananas ever seen in Britain, but he is better remembered for the books he wrote: Volume One of the *Botanical Mercury* in 1634; and Volume Two in 1641. In them he described his plant-hunting expeditions made with other members of the Apothecaries' Society. They called their journeys 'simpling trips' and were as thorough as they were adventurous.

The first was in 1629 to Sheppey and the Isle of Grain, and at once they found themselves in trouble. The local people regarded them with the greatest suspicion and persuaded their constable to lock them up for the public safety. The mayor interrogated them with some care before he would let them go. Then, in the Isle of Grain, a sweltering day without any water or food proved that 'simpling' was no ordinary country ramble. The results, though, of their four days' journey by ship, horse and on foot were very satisfactory: no less than a list of close to 250 discoveries – all early Kentish records.

Johnson's second trip, to the Bay of Margate, was longer, and it yielded over 360 records. His most enterprising piece of 'simpling' though was in Wales. He was the first to botanize there and climb Snowdon in the process, at a time when the most civilized of men still had a real dread of mountains. Even in the following century, his namesake, the illustrious Doctor, when travelling through the Scottish Highlands insisted on the carriage blinds being drawn to shut out the sight of 'horrid crags' without. Apothecary Johnson calmly hired an interpreter and guide – both being necessary in Wales at that time – and, with two companions, scaled Snowdon rooting up plants as they went.

His motive for hunting plants was ostensibly to find new herbs and learn more about the habitats and growth of others. It was professional; though, being an energetic, adventurous young man, Johnson evidently delighted in the trips he described.

The Elm expert, John Goodyer, was a great friend of Johnson's and frequently accompanied him on his expeditions. Being neither an apothecary nor a botanist he must have been one of the first amateur plant hunters in England. Besides his work on Elms it was Goodyer who discovered Ludwigia in Britain and the Lungwort peculiar to the New Forest region, *Pulmonara angustifolia*. His contribution to botany was, in fact, modest; but there was no doubting his enthusiasm for plants. He and Johnson planned to work together on a British flora and make it a comprehensive catalogue of plants with descriptions so clear and scientific that it could be used by medical men who needed to be able to identify plants for themselves. It would have freed medicine from the Mrs Gamps of the age, the 'women who dealt in roots', unscrupulous herb-sellers who, like modern plant nurseries, were not above giving alternatives without authorization. But the grand plan was rendered defunct, and the tyranny of ignorance assured for a few more years, by the Civil War.

This politico-religious struggle was elevated by both sides into a life-and-death struggle for liberty. The two friends preferred that which was offered by Church and King, and Johnson, the younger by eight years, joined the Royalist army – moving from his mild world of hunting plants to the more vigorous pursuit of Round-heads. As a herbalist-apothecary he was an unusual volunteer, but by no means the square peg his superiors might have expected. Undoubtedly his letters to Goodyer would have contained far more information about the plants he found while campaigning than news of the war or his own commissioning by the King and eventual promotion to the rank of lieutenant-colonel. Then the letters stopped coming, and Goodyer had news which knocked the heart out of him. His friend had been in command of the defending garrison at the investment of Basing House in Hampshire.

Basing. It was a name Royalists could remember with pride. Yet the painful, brave story was pathetic, too, because tactically the post had not mattered. The royal position was neither bettered nor worsened by the result of the siege. But a gesture mattered a great deal because morale was always stronger in the Parliamentary army where the pay was fairly regular and the promise of righteous election in the next world never-ending. A noble, if frankly useless gesture might vivify the King's soldiers and so Basing House had been held for as long as possible. It was in the final phase that Thomas Johnson received a ball in the shoulder. The wound corrupted and killed him.

After his friend's death Goodyer abandoned all idea of the

British flora and hardly ever botanized out in the field. His grief was as deep as it was quiet. When the war was over he set himself a monumental time-consuming task, that of writing out the whole of Dioscorides both in Greek and in an English translation. It meant little more to him than lines written out by a punished schoolboy – filled four and a half thousand pages of small and careful hand-writing and took three years to complete.

The Royalist cause found a sympathetic following amongst several noted plant collectors; Robert Morison, for example, who fought at the Battle of the Bridge of Dee and was in exile during the Commonwealth. He returned with Charles II to become Physician Royal, and Professor of Botany at Oxford, and get himself run over and killed by a coach in the Strand. Then there was Sir Thomas Browne who, though his primary interests were natural history, antiquities, and anything quaint or curious, also made collections out in the field. He practised medicine in Norfolk and London throughout the Civil War and at heart was a Royalist.

Thomas Willisel was the only considerable plant hunter to fight on the side of Parliament and it is probably not a coincidence that he served under General Lambert, who was a first class plantsman though not himself a collector. As the discoverer, amongst many other rarities, of the Nottingham Catchfly and the Red Catchfly, Willisel was very well thought of by his contemporaries. They also admired his indefatigableness for he was a hardy, lean man who could live on next to nothing if it was necessary. Many years after the Civil War, Morison, the Cavalier botanist, hired Willisel to hunt out plants in Jamaica, but the soft and deceitful climate leached his strength and, within a year, he was dead.

The attachment of these plant hunters to King or Parliament was the exception rather than the rule. It has been remarked already how many collectors in this era were indifferent to or unaware of the commotions and conflicts which so thoroughly upset everyone else. On the Continent it was also true. Unless they were physically caught up in it, collectors lived through the bloody Thirty Years War, secure in their own religious convictions, and far more bent on botany than battles.

A supreme example in England of the detached plant hunter was John Ray of Essex. He died in 1705, two years before Linnaeus was born, and because of the latter's two-name system of plant nomen-clature Ray's *Catalogue of the Plants of England and the Adjacent*

Islands and the much larger *Synopsis of the British Flora* were quickly made obsolete. This is perhaps why he is less well-known than he deserves.

Ray had much in common with William Turner, the Tudor plant hunter. Both were of humble origin, Turner being a tanner's son, Ray a blacksmith's. Both pulled themselves up by the shoe-strings to Cambridge. Both lost their livelihood because of religious scrupulosity. But whereas Turner was extremely sensitive to the political climate of his day, Ray barely noticed such cataclysmic events as the Civil War and the Glorious Revolution.

While Lieutenant-Colonel Johnson was defending Basing House, the Reverend Mr Ray, a Fellow of Trinity, was pottering about the water meadows by the Cam and making expeditions north to fenland, east to breckland, and south to the knobbly Gog-Magog Hills, collecting and drying plants and working on the local flora. When finances allowed he made long expeditions all over England, and even penetrated well into Wales and the Scottish Borderlands. This he did mostly alone and on horseback, sublimely indifferent to the dispositions of campaigning troops and constant danger from footpads, discharged soldiers and deserters.

The Act of Uniformity after the Restoration drove Ray from Cambridge which, happily for botany, forced him to largen his horizons. He spent three years abroad and then continued his hunting over little-known parts of Britain. The result was a mass of new information and a list of plants which made a mere pygmy of Turner's *New Herbal*.

At this point in his career Ray lost his friend and companion on many botanic expeditions, Francis Willoughby of Windermere, and he chose to retire to the country – to return to his native village, Black Notley. There he married, fathered and raised four daughters and wrote up in elegant Latin the results of his plant hunting over many years. At the conclusion of a successful and, ultimately, tranquil life he died of ulceration. He was seventy-eight years old.

More is known of these collectors of wild plants than of their compatriots who hunted out exotics. But the most petted and fêted, the most lavishly and frequently described, and undoubtedly the most famous plant of the seventeenth century was a tender alien introduced to England long before the Civil War. Affection for the Orange spread like an infection amongst the gentlemen dilettanti of Europe. In England it was all-important in the span of years between the Tudors and the Hanoverians and the culture of over 169 varieties, in pots and tubs on little wheels which were wintered in substantial and often very beautiful orangeries, was the height of fashion and gardening delight.

The fruit was widely championed by people as diverse as the eccentric Celia Fiennes, who rode over ill-made roads and marshy tracks through many English counties, and Sir William Temple, the concocter of that useful word 'sharawagdi'. But it was the scholarly John Evelyn who was the English Orange-grower's cicerone. He advised his friends of the Quality on orangery management, invented his own hothouse stove, an apparatus made of crucible earth pipes and heated with 'a plain single furnace (such as chymists use in their laboratories for common operations)', and published his observations in *Kalendarium Hortense* and *The Art of Gardening*. It is not clear from his diary or his other writings whether or not Evelyn himself collected plants out in the field, but there was no doubting his immense indirect influence on those who did. Moreover, by knowing and loving plants so well and writing of them so tantalizingly, he beguiled people into becoming patrons of exotic plant hunters.

> About Michaelmas (sooner or later as the season directs) the weather fair and by no means foggie, retire your choice greens and rarest plants (being dry) as Orange, Lemons, Indian and Spanish Jasmine, Oleanders, Barba-Jovis, Amomum plin, *Citysus lunatus, Chamalaca trioccos, Cistus ledon clusii,* dates, aloes, sedums, etc.

Nurserymen were quick to meet the demand made by Evelyn's admirers and, having the equipment, primitive forcing houses and orangeries, they cultivated and bred up plants he recommended. There were many of them. In 1648, though it was the climacteric year of the King's final defeat and trial, authorities at the Oxford Physic Garden published a list of the plants grown there. Exactly 1,400, that is two-thirds of the whole collection were exotics – which meant that at some time or another they had been introduced from abroad.

When and why, and who in the first place hunted them out is largely a matter of conjecture. English collectors forced into exile owing to the political and religious intolerance of the age were not uncommon, but it seems certain that the majority of exotics, and undoubtedly those from beyond Europe, found their way into English 'hybernizing houses' by way of diplomatic couriers, merchant seamen and other travellers. Many a sailor's keepsake, brought home as a souvenir of distant lands, had a commercial value. It was in the interests of nurserymen to have a paid agent at the docks whenever shipping made the Port of London. But more often than not plant hunting was a subsidiary to some other occupation.

Spirited adventurers, refugees, social outcasts and misfits were free to make a new and tolerably safe life on the east coast of North America. As they settled and multiplied, driving the native Indians over the Appalachians, the administration of the colonies demanded a supply of clergymen. It was inevitable, too, that in a century engrossed in religion the Anglicans and Protestants of England and the Roman Catholics of Spain and France should send out missionaries to the ungoverned outposts of European civilization and the benighted Indians. Not all, but very many were sufficiently interested in natural history to collect plants and send home seeds and vegetative parts.

Henry Compton, Bishop of London, was one of the most famous collectors of exotic plants in the century. Being born at beautiful Compton Wynates must have seeded his love of gardens and plants though he did not have an opportunity to indulge it fully for many years. As a young man he was a mercenary soldier in Flanders, and on a sudden notion, or divine inspiration, made an abrupt *volte face* of career and returned home at the time of the Restoration to be ordained. Compton's father was Earl of Northampton, high in the favour of the restored sovereign, and his other connections so influential that not many years passed before he was consecrated and enthroned as Lord Bishop of London – with the gardens of Fulham Palace at his disposal. These had already been well-stocked with foreign plants by Bishop Grindal in the reign of Queen Elizabeth, and they delighted Compton. At once he set about the construction of conservatories and the nurturing of plants. Moreover, a chronicler of the day recorded that the new bishop was 'very curious in collecting as well as in cultivating'. He was indeed. 'The Americas' lay within his jurisdiction and he had the responsibility of supplying resident clergymen and itinerant missionaries. On their appointment they were made to understand that an adjunct of their priestly duty to the colonists and Indians was to supply their Lord Bishop with exotic plants. This they did to such good effect that the gardens at Fulham soon contained the largest single collection of exotics in the kingdom. Then, because like most contemporary bishops, he could not keep his finger out of political pies, James II suspended him from all clerical duties – thereby giving him the freedom to enjoy his plants to his heart's content.

One of the best known of Compton's strategically placed clergymen-plant hunters was John Bannister who proselytized for the Church of England in Brazil to small effect, but found there to his bishop's joy the graceful silver-leaved climber named after him, *Bannisteria argyrophylla*. After extensive travels he settled as a parson in Virginia, writing Compton long botanical letters, sending him plants, and working on a flora of the colony. He must be accounted one of the first botanical martyrs because he fell off a cliff while fingering small plants from crevices, and broke his neck.

Loyal always to his botanical clergymen, Bishop Compton saw the Virginian Catalogue through the press, and, busy as he was with this and the palace gardens and conservatories, he was not altogether pleased when James II's suspension was withdrawn and he found himself at liberty to resume his episcopal duties. In fact, the same contemporary chronicler noted that the change in the state of affairs appeared to make little difference to the Bishop's way of life:

> There were few days in the year, till towards the latter part of his life, but he was in his garden, ordering and directing the replacement or removal of his trees and plants.

Happy Bishop Compton.

In concluding these notes on collectors and patrons in the Orangery Era, a place has to be found for the most peculiar of all plant hunters. Like the missionaries, collecting was his secondary occupation. His first occupation, though, was far less godly than theirs.

William Dampier was orphaned as a small child. Thereafter he made life such a hell for everyone that his guardian was delighted to turn a blind eye when the boy ran away to sea. Hydrography began to interest him, and he applied himself to learn as much about it as he could on voyages to Newfoundland, Bantam,

Jamaica and Campeachy Bay. On the rare occasions he stepped ashore he showed an equal interest in botany, collecting specimens for drying, and writing up his observations in a journal. Either in the conflict of interests plants won over water and he skipped his ship in Mexico, or – more likely – he was involved in something disreputable and turned off. In either event he spent two years amongst the foresters of Yucatan, a crew with no laws, no code of behaviour, and no respect for anything or anyone. A willing learner in this school of villainy, young Dampier graduated to piracy. He joined a band of buccaneers who looted and raped and murdered their bloody way across the Isthmus of Darien and southwards down the Colombian coast. Dampier, with his collecting kit as well as cutlass and his readiness to examine a plant on his way to firing a village or desecrating a church, was the living prototype of many fictional villains; of the cut-throats with, say, an appreciation for porcelain, or the torturers with a high regard for family life.

By this time he was acknowledged as a gifted leader and he began his career as a pirate chief by seizing a Danish ship at Sierra Leone. The escapade paid dividends on the voyage he made from Chile to Mexico, thence across the Pacific to China and down to Australia, and it made him the first Englishman to reach the Antipodes. But off the Nicobar Islands his officers and crew decided that he was not such a gifted leader after all, and they marooned him ashore. Evidently he must have had great courage as well as resource for he set out into the frightening vastness of the India Ocean and made his way all alone in a frail native canoe well over two hundred miles to Sumatra.

'Poor Captain Dampier!' exclaimed the Dutch authorities when they had heard his piteous story, and they shipped him back to England.

London lionized him for a time. Well armed, well dressed, wigged and powdered, and with all the charm and salty humour of a good talker and seasoned traveller, 'Captain' Dampier was impressive. Very soon he was being generally admired as the author of an exciting volume called *Voyage Round the World*, and acknowledged by leading plantsmen as the authority on tropical exotics. The Admiralty became interested in such a first-class navigator, and totally ignorant of his buccaneering background the authorities gave him the command of a voyage of discovery to the South Seas where he was to report on the possibility of colonizing Western

Australia. Pepys and Evelyn wined and dined him well and asked him to be so good as to collect plants on their account. He consented.

This second voyage was a remarkable feat of navigation, a necessarily limited but otherwise excellent exploration of the north-west coast of Australia, the coast of New Guinea and New Britain, and the Archipelago and Strait which bear Dampier's name. He had a poor opinion of the Australian aborigines, calling them 'the miserablest people in the world ... and setting aside their shape, they differ but little from brutes'. The flora pleased him far more. Looking for fresh water in the Dampier Archipelago he took the bumboat to an island and, while his men dug for water, he and his water spaniel happily botanized:

> There grow here two or three sorts of shrubs, one just like Rosemary; and therefore I called this Rosemary Island. It grew in great plenty here but had no smell. Some of the other shrubs had blue and yellow flowers and we found two sorts of grain, like beans. The one grew in bushes, the other sort on a sort of creeping vine that runs along the ground, having thick leaves and blossoms like bean blossoms but much larger, of a deep red colour looking very beautiful.

His 'creeping vine' was the Parrotbeak Glory Pea named after him *Clianthus dampieri*. For hunting out this plant alone he deserves fame; though, like him, it is an enigma, being particularly difficult to propagate.*

The expedition almost ended in disaster for the ship foundered off Ascension Island, but Dampier and his crew managed to get ashore where they lived off turtle and goat until they were seen and rescued two months later. In the shipwreck Dampier lost most of his collections and his notes and sketches, but 40 per cent of his

---

* The compilers of some gardening dictionaries are sanguine to say that cuttings root well in sandy soil under glass, or that the plant can easily be propagated from seeds. The experience of weathered gardeners is generally to the contrary. The Parrotbeak is obstinate. It requires pampering. It must be constantly nourished and invigorated by a more robust shrub. This means sowing seeds of the Bladder Senna in thumb pots, and two weeks afterwards, sowing the Parrotbeaks. When the latter are still minute, i.e., before they have formed their first true leaves, they have to be grafted on to the Bladder Sennas. If they take well, thereafter they will thrive, but the actual grafting is a tricky job requiring the deft and little fingers of a Japanese or an expert chicken-sexer.

dried plants survived and can still be seen today in the Oxford Herbarium.

By this time the strain of being a respectable sea-going captain was proving too much for Dampier. He began to vent his spleen on the nearest person to hand, his first lieutenant, and was so atrociously cruel to the poor man that when they made port back in England he was denounced by the crew and court-martialled.

To be tried and found guilty of cruelty in the Royal Navy at a time when, in Winston Churchill's memorable phrase, 'rum, buggery and the lash' were keeping her together and beginning to make her great, was in itself remarkable. Then, only three years after being convicted, to be re-appointed to an even higher command in exactly the same navy, was even more so. But this is what happened.

In his new venture to the South Seas Dampier had overall command of two ships. The captain of one (improbably enough) was a London physician and the inventor of a sweat-producing sedative called 'Dover's Powder'. And the master of the same ship was Alexander Selkirk, a scamp from Fife who fell out with the doctor and, at his own request, was put ashore on Juan Fernandez to become the prototype of Robinson Crusoe. On the voyage Dampier's navigation was superb, and his plant hunting highly successful, but his old devil outed again. This time when his little fleet reached home port again he was accused of excessive brutality and drunkenness and even the fruit of both, cowardice. He wrote an angry *Vindication* but it made no difference. His reputation was gone, and he was unemployable as a commander. In view of this it is surprising that he should have found his way into *Gulliver's Travels*, and that as 'a rough sailor but a man of exquisite mind' he should later have inspired Coleridge to write *The Ancient Mariner*. But there could be no satisfaction in this for Dampier. For the present he was an ex-officer and thoroughly vexed and depressed, until adventure and travel and the wild plants of the South Seas re-exercised their spell over him. As a result when, precisely four years, four months, and four days after his marooning, Alexander Selkirk was taken off his island by a privateer, he found that the navigating officer was none other than his old commander and that bloodthirsty botanist, William Dampier.

4

The Golden Age of Botany

It has been suggested that in the seventeenth century plant hunting was mostly motivated by the need to know more of medicine, and then by the desire to find either interesting native and exotic 'greens' or suitable parterre flowers for the *broderies* of geometric gardens. In the following century plants were primarily collected in a spirit of scientific inquiry. The botanist's daybook and herbarium, his microscope and vascula became more important than the apothecary's laboratory and the garden bed; and from this point in time the field widens so far (and continues to widen up to the present day) that no more than a cursory history of hunters *en masse*, or thumb-nail sketches of a representative few, is in any way possible. Deciding what material shall be included in a book is often less difficult than deciding what shall be left out. In this study the opposite is true; but, metaphorically throwing all the principal botanical collectors of the age into a sieve, three will not drop through with the rest. They were almost exact contemporaries as well, all dying in the 1770s: Linnaeus of Sweden, Commerson of the Dombes, and John Bartram of Philadelphia. They are selected as representative, no more; which is not to deny the historic importance and the influence of the botanists, plantsmen and hunters from other backgrounds and from other places, say, Virginia or Vienna, Potsdam or Peking.

*

The eighteenth century has been called the Age of Reason, despite the sublime superstitiousness of the time; and the Age of Elegance, which appeared to pay small regard to the reek and the encrusted dirt beneath the scents and silks. With complete justification it has also been called the Golden Age of Botany for it was the century of classification, of systematic plant hunting, of professionalism, and, within degree, of experiment.

A good deal of this was due to the work of Carl Linnaeus.

The number of known plants increased each year. By the date of his birth in 1705 the local and regional and national and international floras – based on Dioscorides–Aristotle or on personal observation out in the field – had grown top heavy. There were also competing systems of classifying plants which, with the over-long names and descriptions, and the meretricious, herbalist clap-trap which clung to plants down the centuries, resulted in a jungle of muddle and mistaken identities. Some-one with a clear mind and a keen eye was required either to clean up the mess or to abandon it and suggest an acceptable alternative.

Linnaeus* was the man.

His father was the pastor of a country village in the south of Sweden. As frequently happens with sons of the parsonage, he was destined for the ordained ministry of the Church, but – like many another parson's son – he managed to evade it. In fact this was not particularly difficult. From the time he was able to run Linnaeus showed such a liking for being out in the fields and such a loathing for any sort of schooling that his father was seriously advised to apprentice him to a tailor or a cobbler. The local doctor checked this by taking him into his own house and encouraging him to study medicine. So, when the boy went up to the University of Lund he went to read medicine not theology.

Linnaeus's year at Lund and the following years at the larger University of Uppsala were not easy. He was a poor country boy without even the money to have his shoes mended. They were patched, amateurly, with folded paper. And he was often hungry. But at both universities he was befriended and looked after by men in comfortable circumstances: at Lund by the amateur botanist Doctor Stoboeus, and at Uppsala by a Doctor Olaf Celsius, priest

---

* It is convenient to refer to him by his latinized surname, though Carl Linné was his real name and, in his fiftieth year he was ennobled and given the sought-after 'von'.

and botanist, who chanced upon him in the academical garden,*
and promptly took him home, gave him board and lodging and the
use of his extensive library. As a token of his gratitude Linnaeus
tutored the doctor's son in botany, and then, when he discovered
he had an aptitude for teaching, his fellow undergraduates. Quite
soon he showed such promise that the university authorities
appointed him assistant to the Professor of Botany and, even before
he could settle to the immensely hard work and minute income of
this post, the Academy of Science elected him to explore and hunt
plants in Lapland. He had not yet taken his degree, but there
would be time later, and he was young and the offer was a marvellous one. He accepted the election with delight and set off on a
journey of hundreds of miles through unmapped territory writing
up his journal each day, trying to develop his embryonic ideas on
classification and making a collection of dried plants.

It was 1732, a time when people dressed according to their class
and wore the same kind of clothes whether they were playing football, gardening, or attending an execution. It did not make for
great comfort though the young Linnaeus appeared not to notice
this:

> My clothes consisted of a light coat of Westgothland linsey-
> woolsey cloth without folds, lined with red shaloon, having
> small cuffs and a collar of shag, leather breeches, a round wig, a
> green leather cap, and a pair of half boots.

His equipment was as light as he could make it: a shortsword,
a fowling piece, a pocket-book and passport and a little leather bag
which contained:

> One shirt, two pairs of false cuffs, two half shirts, an inkstand,
> pencase, microscope and spying-glass, a gauze cap to protect me
> occasionally from the gnats, a comb, my journal, and a parcel of
> paper stitched together for drying plants.

False cuffs seem an unusual requirement for a plant hunter in a
largely uninhabited area.

It was an exhilarating trip. Linnaeus returned home with a

---

* It is singular that three of the most famous plantsmen have been
chanced upon in botanic gardens and promptly befriended: Linnaeus by
Dr Celsius, David Douglas by Sir William Hooker, and Joseph Paxton
by the Sixth Duke of Devonshire.

hundred dried plants, hitherto unknown, to establish his reputation as the author of the *Flora Lapponica*, and fall in love with a sweet young girl who adored him in return.

The sweet young girl's father, though, would not hear of her marrying a penniless botanical collector. Linnaeus must take his degree, practise as a physician, and settle down. The father was reasonable enough to provide the necessary expenses, and as the money was available Linnaeus sensibly decided to see a little more of the world and go to a foreign university.

While plant hunting his way through Northern Europe the young man's zeal for scientific accuracy caused him to be heartily disliked in Hamburg. There the city fathers were custodians of an ancient and prized 'seven-headed hydra', a relic of the Crusades to the Holy Land. Linnaeus, as a humble student, asked for permission to see the marvel which he then examined with care. His thoroughness established that seven weasel heads had been skilfully sewn together, attached to two clawed feet, and covered with snakeskin. The triumph of scientific observation entirely failed to please the city fathers who up to that time had enjoyed a small but regular income by putting the creature on show, and Linnaeus wisely hurried on his way.

He took his degree in Holland, and for two years was Curator of the Hatrecamp Botanic Garden. For a long time he had been perfecting his system of plant classification, which had its foundations in Sébastian Vaillant's observations on the precise sexual functioning of plants, and at last he published his theories.

They were widely acclaimed because his approach was so unusual and commendably simple; his system asking no more from a botanist than that he count the number of a flower's stamens. The total, noting certain other characteristics, would place the plant first in a division and then in a sub-division. Identification was as easy as that.

But it had its faults as such an artificial system must have,* and though explicable in Linnaeus's Latin original, it was open to bizarre interpretation in foreign translations. In one late eighteenth-century Englished version, for example, endless complications and ludicrous situations arose through calling the stamens 'husbands' and the styles 'wives'. This caused one plant to have

* Nevertheless stamen-classification remained the dominant system for a long time.

'twenty husbands or more in the same marriage'; and another plant 'four husbands, two taller than the other two'. As a result there was a farcical outcry in Wesley's England against the Swede's lewd and salacious botany.

Linnaeus's most enduring contribution to botany was his formalizing of Caspar Bauhin's idea of giving two names to all created things. He undertook to name and classify everything 'from buffaloes to buttercups'; and undoubtedly his work on indexing plants published later in his life in 1753 as *Species Plantarum*, made something orderly out of the casual, chaotic botany of preceding years, and stabilized nomenclature for ever.

The young Swede's reputation was already considerable. Because the owner of the Hatrecamp Botanic Garden was an English East Indian nabob it was natural that Linnaeus, as Director, should visit England. When he did he was touched by the warm welcome he received from many of the distinguished botanists of London. Fame pleased him, and recalling that prophets on the whole are not known in their own country, he might well have decided to stay away from Sweden. But he was still young and the sweet young girl was awaiting him in Stockholm. In the event when he did return to marry and practise medicine he was gratified to find himself as well known in Uppsala as he had been in Western Europe. But neither marriage nor doctoring suited him. The sweet young girl turned almost overnight into a dragon. All his life he was to regard her 'with respect and terror'. And as to being a physician, he expressed himself truthfully (but clumsily, it is to be hoped) when he said that he was 'fonder of meddling with plants than with patients'.

Liking neither wife nor work he considered improving the latter by leaving Sweden but, fortuitously and just in time, he was elected Professor of Anatomy at the University of Uppsala. He did not care for the subject but he accepted the chair and it was a step in the right direction; that is, towards his real vocation as a teacher. Moreover, not very long after he was able to exchange his professorship for the less lucrative but far more congenial chair of Botany, Materia Medica and Natural History.

Teaching, sharing his own pleasures, and enjoying the company of adoring students was to fill the rest of Linnaeus's life and make him supremely happy. His influence was huge. At the beginning of his teaching career as professor, Uppsala had 500 registered undergraduates. The number gradually increased to three times that number. Students came to him from all over Europe. At his death, the number dropped once more to 500.

Included amongst his followers were some who were to become leaders in the natural sciences throughout the world. People of distinction from bishops to leading botanists would travel to Uppsala and attach themselves as extra-mural students for the week-end, simply to enjoy the privilege of hearing Linnaeus lecture.

The amount of enthusiasm he generated was astounding and, when the malignant Swedish weather allowed, his twice-weekly botanical expeditions would attract as many as 200 undergraduates and more; often so many that they announced fresh discoveries to one another with bugle calls like huntsmen in the field. At the end of the hunt they took their trophies to the professor to hear him appraise the best of them. If he judged any to be of particular interest they would honour the plant and its discoverer with a celebration – singing as they marched home in a long crocodile to the beat of drums and fanfares from bugles and trumpets.

The party that evening would be held in a hall or garden with much dancing to fiddle music and drinking as only Scandinavian students could drink. Their professor would sit smoking his pipe and enjoying their exuberance. And then he, as well, would be pulled to his feet to share in the dancing. At this, too, he excelled. Not only was he an internationally renowned botanical collector and classifier and their inspiring teacher, but he could also outclass them all in the footwork of a Polish Jig.

*

In contrast to the social plant hunts of Linnaeus's later years – when legions of students carried the vascula and presses and all the paraphernalia of collecting, and would, had it been necessary, have carried their professor too – John Bartram of Philadelphia chose to be as solitary as possible in forty years of plant hunting along the eastern colonies of North America.

Legends were told about this man in his own lifetime. It was partly because he was so reckless in penetrating wild Indian country at a period when the Pennsylvanian communities dreaded the Iroquois away to their north and the Chickasaws and Shawnees to the west of the Alleghany Mountains. But it was also because the *mise en scène* of Bartram's whole life was intriguing, to say the least.

He was the son of a Quaker English farmer who reached the colonies in 1680, 'before there was a single house in Philadelphia'. Old Bartram was a single-minded individualist. On the death of his first wife (John's mother) he handed his children over to their grandmother, quarrelled with the local Friends, defied the 'Meeting' when it inquired into his personal affairs and promptly emigrated to North Carolina. There he remarried, bred two more children and ended his rumbustious life speared and scalped by Yamasee Indians beside the River of Fear.

John Bartram was as unable to escape polemics as his father. He, too, was hailed before the 'Meeting' (which appeared to have the powers and the charity of a soviet) and told to account for his peculiar views. Like his father he denied their right to pry into his private affairs. Instead of walking out, though, he remained where he was while the 'Meeting' drew up a final 'Letter of Irrevocable Disownment', and he politely took his seat each Sunday in the Meeting House as if nothing at all had happened. The perplexed and embarrassed Friends did not quite know what to do in the face of such eccentric behaviour, so they did nothing.

The average colonist was a farmer and John was no exception. Bred up to ordinary husbandry he showed a talent for imaginative farming which was far in advance of his time, regularly feeding his fields with a rich liquid manure, spreading rotted compost over virgin and impoverished land during the winter months. This gave him lush hay 'outside a donkey's dreaming', and he drew an average of thirty-three bushels of wheat from an acre while his neighbours averaged only twenty.

He needed to do well for he had a large family. His first wife died in 1727 after four years of marriage, leaving him a widower with two small children. A year later he bought a tract of land on the banks of the Schuylkill River which edged on the wilderness. This he farmed. He threw together a timber shack, remarried, built a larger house round the original, and added to it as ten children were born of the marriage. One of them, William, was to be as keen a botanist as his father and, towards the end of his life, accompany him on some of his plant-hunting expeditions.

How, in the first place, John became interested in plants has been the source of two romantic stories. The first was that he found a colony of Birdfoot Violets, *Viola pedata*, in a water meadow, admired the flower so much that he dreamed of it the following night and woke at dawn a perfervid devotee of the science of botany. The second, told by his French friend, the scientist Crèvecoeur, had the same basis of sudden revelation, but the cause, in this case, was a single Daisy. Bartram was ploughing, saw the Daisy, plucked it, pulled it to bits out of curiosity, and then (according to Crèvecoeur) chided himself in the following Quaker-ish fashion:

> What a shame that thee should have been employed so many years tilling the earth and destroying so many flowers and plants, without being acquainted with their structure and uses.

Which remarkable rebuke caused him to hire a man to continue with the ploughing while he himself went into Philadelphia to buy a book on botany and a Latin grammar to make it explicable.

Later in his life Bartram made it clear to his London friend and patron, Peter Collinson, that from the age of ten he had been mildly interested in plants; and that, as the years passed, the interest expanded until it grew beyond his control and became an obsession.

The second Mrs Bartram did not care for this at all. She had married a farmer with prospects, and there he was, studying and collecting unprofitable weeds, and hiring a hand to do the farmwork with money they could scarcely afford. To begin with, their life was indeed difficult, and Bartram had to temper his enthusiasm in order to meet the needs of his farm and ever-growing family. Not being able to get out and hunt plants in the wilderness as often as he wished, it was at this time that he personally laid the foundations of his famous five-acre botanic garden. Then good luck, nothing else, put him in touch with Peter Collinson.

Collinson was a London merchant who traded with the American colonies. He was also an amateur botanist and plantsman who had a wide circle of like-minded friends in England. From the plant introductions made by young John Tradescant and, since then, by traders and missionaries in the colonies, he estimated that the flora of the Americas must be fascinatingly different from the flora he already knew. He tried his hardest to persuade settlers to send him supplies of plants. Some obliged but the majority of his friends in the colonies shared neither his interest nor his knowledge. Even those who were willing and able to collect for him could only be asked once, or twice at the most. To pester friends for consignments of plants would be ungracious. Collinson plagued his business agents to find him a professional plant hunter. It was outside their job and their attempts were only half-hearted. Then good luck, nothing else, put Collinson in touch with John Bartram.

To begin with they were shy. Neither wished to embarrass the other and jeopardize the future. One was delighted to pay five guineas a box for plants and seeds, and the other equally delighted to have a reasonable excuse to indulge in his passion; but, in the first place, they were modest in their hopes and demands – Collinson not daring to hope Bartram would agree to send a second or a third box; Bartram not daring to hope that Collinson would wish for a second or third box. In the event, there was no question of second or third boxes. For thirty-three years John Bartram hunted American plants for Peter Collinson, sending hundreds of boxes on the long journey across the Atlantic. During the Seven Years War some consignments were lost, and even in peacetime only a small percentage of plants could be expected to reach England in good condition, exposed as they were to dangers from weather, salt spray, and the understandable pococurantism of seamen. Vermin added to the dangers. Collinson once found a rat's nest, complete with young, in one of his precious boxes, all of them well and apparently content 'amidst a ruin of roots and dead greenery'. But the percentage of successes was large enough to make the enterprise worth while.

Collinson gathered together a syndicate of botanists and plantsmen willing to pay an annual subscription of £10 for a share in Bartram's boxes. And he also found private patrons for his protégé in Pennsylvania. The most considerable of them was Lord Petre, who would underwrite the cost of a complete expedition, but their number also included the Dukes of Richmond and Norfolk, and the Earl of Bute, one-time Prime Minister. Bute became a martyr to collecting when he was out botanizing, fell out of a tree and killed himself. He also hired Bartram on behalf of Frederick, Prince of Wales – better known as the 'Poor Fred', who died in the arms of his dancing-master. With such influential patrons it is not surprising that Bartram's highly successful plant hunting was officially recognized, and Collinson was able to write him:

My repeated solicitations have not been in vain, for this day I
received certain intelligence from Our Gracious King that he
had appointed you his Botanist, with a salary of £50 a year.

As Collinson's protégé, and in particular as King's Botanist,
Bartram was in the position to do much as he wished. Though until
his boys grew up to take his place he stayed on the farm at sowing
time and harvest, his interest and most of his energy was reserved
for plant hunting in the virgin American forests, plains and
mountains.

In scores of places he was the first white man to penetrate into
the interior, though on occasions he came across 'White Indians',
that is, men and women who had been captured by Indians, and
reconciled to their customs, preferring them to the white way of
life. Bartram did not like them. He was notoriously un-Quakerlike
in his illiberal attitude to the Indians and the Whites who had 'gone
native', though as the son of an Indian victim it was understandable. But there was less language problem with them than with
real Indians and sometimes he used them as guides. On Bartram's
trip to the Great Lakes, for example, one of his guides was called
Shicckalamy and professed to be chief in a village of Delaware
Indians, though he was in fact a Frenchman from Montreal. His
interpreter on the same expedition was Conrad Weisser, a refugee
from the Palatinate who had tried his hand at being a recluse in the
Seventh Day Cloister at Ephrata, but far preferred belonging to
the Maque nation of Indians. Bartram called them savages and
'gone wild Whites', but they were useful.

Generally though, when hunting plants he was alone, and
because he was often in totally unexplored country and used
his own names to identify geographical features, it is not
always possible to discover precisely where he went. His maps
were hypothetical and most of his ringing, descriptive names like
the Impenetrable and Endless Mountains and the Dismal
Vale were never adopted. None the less it can be safely taken
that he travelled widely even during the Seven Years War
against France and that he went deeply into Indian country as
well.

He carried little with him; a bedding roll, a change of linen,
books, tools and firearms, and two panniers attached to his saddle
for plants and seeds. He took salt and flour as well, but otherwise

lived off the country, eating berries, nuts, fruit, edible stems, the game which he shot and the fish which he caught. If the Indians were friendly he would eat with them, confiding to his journal his high opinion of their cooking. Fresh eels cooked over embers and eaten with boiled corn and water melons won his special commendation; and so did a bowl of boiled beans mixed with Indian dumplings made of new and soft corn scraped from the ear. Accustomed to frugal and exceedingly plain living at home he grew quite excited about the Indians' capacity to make a feast of Wake Robins. And a banquet of wild Huckleberries and Maize seethed in venison broth drew from him the equivalent of a Michelin gastronomic star:

'Noble entertainment. Too good to leave any at all.'

Bartram has been credited with introducing more than 200 American 'discoveries' to the gardens of the Old World. They included the Magnolia, *M. grandiflora*, which was exceedingly popular and became 'as essential to a fine Georgian house as the walls that supported it', and *Lilium superbum* which, when it first flowered in Peter Collinson's garden in 1738, threw him into such a state of enchantment that his family and the servants were seriously alarmed.

Patron and protégé became fast friends and regularly corresponded although the mails were so slow that a large packet of numbered letters would be delivered instead of a single communication. Bartram was a great reader and longed for books from London. Although it was costly Collinson did his best to oblige him, but when it became evident that Bartram's appetite was insatiable, he sent a mild note warning him that even Solomon did not get all his knowledge from books. Bartram replied wryly:

I love reading dearly, and I believe if Solomon had loved women less and books more, he would have been a wiser and happier man.

The King's Botanist of Philadelphia is justly regarded with pride in the United States. The five-acre botanic garden by the Schuylkill River which he called his 'garden of delight' is mostly extant as a part of the present-day Philadelphia park system. His most enduring memorial, though, is the accolade bestowed upon him by Linnaeus himself. In admiration for his immense, self-taught knowledge and his assiduity in hunting plants, the great classifier called John Bartram 'the greatest living botanist in the world.'

*

Linnaeus, in his *Glory of the Scientist* of 1737 exclaimed:

Good God! when I consider the melancholy fate of so many of [botany's] votaries I am tempted to ask whether men are in their right minds who so desperately risk life and everything else through their love of collecting plants.

Unknown to the Swedish savant he had written an apt description of his admiring disciple, Philibert de Commerson. Certainly Commerson thought so when he read it, shyly admitting to a friend that he had what he called 'botanomania'. He had indeed. Pasfield Oliver in his life of Commerson* underlined the point:

The mere suspicion of a plant unknown to him was an irresistible attraction. He thought nothing of scaling almost inaccessible mountains, and both risked his life and ruined his health by his exertions . . . On his botanical expeditions he always went by himself, taking with him scarcely any money or provisions. He would return home sick, scarred with wounds, shaken by falls and accidents of all kinds, as well as utterly worn out by his exertions and by the violence of his own enthusiasm.

There was always a luminous air of unreality about this vital, frenziedly energetic and prodigious tornado of a plant hunter. Even his beginnings were rooted in fantasy for he was a Franc Bourguignon, born in the Dombes, then an independent state lying between the rivers Ain and Saône, comparable to those present-day curiosities, Monaco, Andorra and San Remo. In theory the Comte d'Eu was sovereign prince; in fact the place was outside any legal authority. There was no security for life or property except 'the ancient habits and customs of the Dombes' until France first purchased and afterwards formally annexed the territory into the kingdom in 1762.

Commerson was born in this strange place in 1727, when the country was infamous as an unhealthy seat of agues and rheumatism and a species of low malarial fever. Though a plateau, it lay so low that it was perpetually dank, and, being a desolate, uninviting, unwanted tract of shallow valleys, coppices and stagnant

* *The Life of Philibert Commerson* by Pasfield Oliver, edited by G. F. Scott Elliott (John Murray, 1909).

pools, it was a refuge for hundreds of thousands of waterfowl, some of them extremely rare, and the habitat, as well, of many unusual plants. Had he chosen it, Philibert Commerson's bizarre birthplace could hardly have been more appropriate for an incipient naturalist.

His first love was for fishes, which in such a sodden environment was barely to be wondered at and particularly so as the Dombists practised an unusual fish husbandry. Over roughly 20,000 acres they rotated fish and corn as commercial crops. Lakes and ponds were drained, the fish taken, and grain sown on the muddy bottoms. After harvest they were refilled and restocked with fish. These fattened on the gleanings of grain, provided ready food for the Dombists and manured the mud which, in two years' time would once more be a field of growing corn. This extraordinary rotation of carp, tench and other fishes with oats and barley gave young Philibert a special interest in fishes. He collected them with gusto and conceived a highly original idea for preserving them; that is, by drying the little ones between sheets of heavy-gauge paper just as though they were marsh plants. But the boy was fickle in his affections. From fishes, birds and birds' eggs, even from the fascination of reptiles and fossils, and the cohorts of brightly coloured insects, the plant world stole his heart away.

Though he had his quiverful of fourteen children Philibert's father was rich enough to set them all up. Seeing his second son's obsession with botany he sent him to Montpellier University, not as large as some, and in a draughty part of France, but the nursery of l'Obel, l'Ecluse and many other botanists of stature. It had, moreover, a botanic garden of some distinction and so, evidently, it was exactly the right place for Philibert.

It would have been if the boy's absorption in plant hunting had not led him to forget there were such things as petty larceny and the law of trespass. It was, and remains a common practice amongst many plantsmen to take cuttings and filch bulbs and seeds, even whole plants. If outright purchase is impossible and proprietors will not be persuaded to part with plants looting is the only solution. Commerson had precedents for his behaviour. Joseph Pitton de Tournefort, the adventurous French botanical collector of the previous century who in a plant-hunting expedition to the Levant collected 1,300 specimens, was given to looting the gardens of his friends, and sometimes those of total strangers. On one occasion he was caught in the act by an outraged gardener who bombarded him with a fusillade of rocks and brickbats, and with such devastating accuracy that Tournefort, who was on terms of intimacy with the king and had thought nothing of plant hunting through thirty-two Aegean Islands in stormy and pirate-infested seas, was obliged 'to save his life by extreme agility and the fleetness of his feet'.

Tournefort was deeply admired by Commerson and faithfully imitated. It was bad enough that very soon the young man was in trouble with a large number of smallholders round Montpellier, but worse when he was actually caught collecting specimens for his herbarium in the University Botanic Garden – and not merely by a gardener or keeper but by François Boissier de Sauvages de la Croix, the Professor of Botany himself.

It was a sign of the advanced state of Commerson's botanomania that he was outraged and truly considered himself an injured party when he was barred from the garden. The interdict was only temporary but he never forgot it, and years afterwards he was still criticizing the Professor who had caught him helping himself to university plants. In botanical circles he was quickly recognized as a brilliant but apparently lunatic plant hunter; and it is clear that between leaving the university and his marriage in 1760, he became almost a compulsive collector – a magpie of living plants, herbarium specimens, botanical treatises, books and catalogues, in fact any absorbable piece of information. The risks he took for the sake of plants were quite extraordinary, and if he had not been exceedingly sharp-witted, athletic and lucky he must, surely, have succumbed to one or other of them. On one dramatic occasion he heard the roar of an avalanche above him and, not waiting an instant, for it was imperative to beat the avalanche in the gravity race, he hunched his knees to his chin and rolled like a ball down a fiercely steep gradient to safety. On another occasion he was collecting along the steep banks of a swollen mountain stream when he slipped and was caught like Absalom by the hair in a bush. To free himself 'he was obliged to cut his hair and fell into the stream by which he was carried away and very nearly drowned'.

Marriage calmed him. His wife brought him fortune and gave him a much-loved son, Archambault, but two years later she died.

Grief stirred Philibert up again. He became determined to be a 'practical' botanist, and not shy away from the hardships and the hazards which hunting plants involved. The study, the microscope and the herbarium were not sufficient for a botanist like Commerson. 'Glory,' he wrote, 'requires, like Fortune, a race of men alike hardy and tenacious.'

Chance gave him the opportunity to gain and deserve just such glory: the King appointed him Royal Botanist and sent him off on a circumnavigation with that prodigy of many arts and sciences Louis Antoine de Bougainville. Besides being a favourite of the King, and described by a contemporary as '*plus orné, plus dix-huitième siècle*,' Bougainville was also widely admired as an advocate at the Paris Bar, a diplomatist, a mathematical genius and the author of *A treatise on the Integral Calculus*, a Fellow of the Royal Society of London, an eminent soldier and navigator, explorer and linguist, and so sympathetic in his understanding of the North American Indians that he was joyfully accepted into the Tribe of the Tortoise. Moreover, notwithstanding his friendship with the French royal family, Bougainville survived the terror to serve France under Bonaparte; and this he did in no sense as a Gallic 'Vicar of Bray', but with his honour and integrity intact. It is ironic, perhaps, that so noble and brilliant a pantologist should be best remembered today because of the exotic plant named after him, the Bougainvillaea. Both having such an exuberant taste for scientific inquiry the expedition's commander and the King's Botanist were bound to get on well.

Before he left, Commerson handed over his son Archambault to the care of an uncle and solemnly made his will. It was an enlightened document in which he ordered that if he should die in a place where there was a medical school only his heart should be buried and the remainder presented to the nearest theatre of anatomy. He also 'attached in perpetuity the capital and interest of two laundries near Châtillon' for the founding of 'a prize of morality' – whatever that might have been. He made several bequests, provided for his housekeeper, Jeanne Baret, and left the rest to Archambault.

Then he raced to Rochefort, extravagantly changing horses as often as possible, and in an agony of suspense all the time because he was a little late. He need not have troubled. His ship, appropriately named *La Boudeuse*, kept to her moorings for several days after Commerson's breathless arrival. This come-down, though vexing, gave him the opportunity to stow and arrange all the scientific equipment with Jean Baret his young botanical assistant.

When they did sail Philibert found he was a poor sailor. The heave and the rock of his new home made journal-writing difficult. A brief entry headed '4th–7th February' makes everything plain:

My dinner and supper are but loans, which I am exact in repaying one hour or half an hour afterwards.

The French King's commands had been loosely phrased – 'to proceed to the East Indies and so circumnavigate the globe'. The idea was to achieve for the glory of France what had already been achieved by the English, and – in passing – to claim and name pieces of unexplored territory. The west coast of New Holland, as Australia was then called, had already been partially explored and claimed. So far as anyone knew, the east coast had not. At a later date the French venture was charmingly described by its commander as 'Bougainville's Failure to Discover New Holland'. In fact, like Scott of the Antarctic, he was just a little late. Territorial considerations aside his circumnavigation was a triumph of seamanship and a great success especially in the fields of anthropology and natural history.

A list of Commerson's plant discoveries alone would demonstrate his outstanding contribution to the expedition. There were at least 60 new genera and 3,000 new species. But as naturalist to the expedition his work was by no means confined to botany.

His journal and the ship's log show how versatile were his activities and interests.

When, as an illustration, they reached the Rio de la Plata he saw at once why the Spanish explorers of 1513 had named it the River of Silver, for billions of white Zephyr Lilies (*Zephyranthus candida*) were in bloom and the plains gleamed with a silvery sheen. The sight proved to be but an appetizer. The exquisite beauty of Brazil delighted him. 'I had a fearful sore on my leg which appeared at sea,' he noted, but it did not prevent him from botanizing and, amongst other major discoveries, he found a pink species of the Zephyr Lily which was named for him *Zephyranthus commersonii*.

Afterwards, in Buenos Aires, he eschewed botany for long enough to hunt jaguar with two gentlemen adventurers, a German prince and a French chevalier who had paid to join the expedition.

Then, still farther south, on the littoral of the Bahia Grande, where the cold was so devastating that the hardiest seamen wore all they possessed and still were chilled to the bone, Commerson and his assistant hunted a less ferocious but not less elusive quarry, adding handsomely to their collection of discoveries. Jean Baret earned his master's warm praise on this expedition for his fortitude, his ingenious attempts to ameliorate their discomforts, and for his willingness to carry so large a share of the firearms, forks, vascula, provisions, drying papers and presses, sketching blocks and other equipment.

The snow-water they took on board at the Straits of Magellan gave everyone sore throats, and it was Commerson the naturalist and not the ship's surgeon who provided a remedy: ordering that each barrel of water should be laced with a pint of vinegar and mulled with red-hot bullets.

On the strength of this original cure they made Tierra del Fuego, and there the ship's blue-grey cat which had attached itself to Commerson so captivated the natives that they offered him one of their loveliest young girls in exchange. Cat worshippers will be gratified to hear that, under the circumstances, he felt bound to refuse.

Like countless ships before and since, *La Boudeuse* appeared to hesitate in apprehension before facing the Horn. But Bougainville's confidence was caught by everyone on board. His seamanship and navigation were never in doubt. As calmly as if he were promenading in the Bois de Boulogne he carried them safely through that region of icy peaks and screaming, shrieking, never-dying winds. Then they breasted the Peru Current and made good way, all sails ballooned by the South-East Trades.

Commerson's absorbed interest in natural history was again temporarily suspended by sea-sickness. But he was certainly in trim and game for everything when they made landfall at Tahiti where the prototype of Rousseau's 'noble savage' eked out his paradisical al fresco existence with public love-making, fruit- and pig-eating, and, on occasions (in case there should be vengeful gods), with the excitement of a human sacrifice. Commerson confided to his journal a conviction that in fact the Tahitians knew no god but love. The distractions from botany were formidable but he did manage a little plant hunting in the thirteen days of their stay. At sea again his time was partly filled with examining marine life and pressing and preserving small fishes as he had done in the Dombes as a boy. Always on the alert for scraps of miscellaneous information he noted in April 1768:

> 16° s. lat. & 170′ e. long. Gold-tailed Dolphin affords excellent eating when cooked with butter and capers.

Commerson's even, contented life suffered an abrupt upheaval on Mallicollo Island in the New Hebrides when his shipmates were made aware of a singular fact.

A local chieftain, admiring Jean Baret and free from Western inhibitions, seized the young botanist and would have made off with him if a shore party of seamen had not prevented it. In the scurry Baret's clothing was somewhat disarranged.

Bougainville noted the facts with dignity and detachment:

> M. de Commerson's servant (Baré by name) . . . confessed to me her eyes streaming with tears, that she was a woman. She told me that at Rochefort she had deceived her master by presenting herself in men's clothes at the very moment when he was about to embark.

Pasfield Oliver, writing for Edwardian readers, assumed a cautious tone:*

> It is best, after the lapse of one hundred and sixty years or so, to add no comment whatever to this extraordinary story.

More recent research and a puckish irreverence to the subjects of thumbnail biographies has suggested there is a close resemblance between 'Jean Baret' (or Baré) and the housekeeper 'Jeanne Baret' mentioned in Commerson's will; and has provoked, as well, the legitimate question: Was the botanist naïve, wily, or frankly stupid?

One delightfully unlikely story has it that, in consequence of the discovery, poor Commerson, being old (he was forty-one!) and much mortified (and he an eighteenth-century Frenchman by no means averse from the pleasures of the alcove!) was so embarrassed that he decided to exile himself for ever; and that Jean changed her

---

* At least in refraining from comment and not concealing facts he was more honest then the overfastidious Victorian 'editors' of plant hunters' journals. Joseph Hooker, for example, bowdlerized Banks's robust account of his South Sea voyage in search of plants to such an extent that a modern authority has described his work as 'not so much a journal as a piece of carnage'.

name to Hortense, stayed to comfort her master until his death, when she went home to Paris, married a clockmaker, and gave her name to the commonest garden Hydrangea, *H. hortensis*.

Pull, as they say, the other one: nevertheless it is the best piece of whimsy attached to the career of any plant hunter.

It seems that, in fact, Jean rather naturally changed her name to Jeanne, that she married a soldier, and that she returned to France where she settled near Commerson's family in the Dombes. At her death she left everything to Commerson's son Archambault. Commerson, in his turn, honoured her by naming a genus *Baretia*, though a prior name being established it is now altered to *Quivisia*.

Clearly the naturalist and his assistant meant a good deal to each other, and Commerson never did go home. The reason for this is clouded. He left Bougainville, and presumably Jeanne Baret, at Mauritius, and fetched up in Madagascar. Promptly he set about the flora of the island, and to such effect that he had it pigeon-holed in record time. Then he was recalled to France, his appointment cancelled and his salary stopped. This, anyway, was how he put it when he wrote a vexed letter to a friend in Paris, exclaiming: 'My plants, my beloved plants have consoled me for everything.' But, probably, the communication from Paris was neither as brutal nor as mysterious as it appeared, bearing the reasonable interpretation that Commerson's appointment and stipend would inevitably have finished when he left *La Boudeuse*, and that he was bidden home again in order to reclaim the post of Botanist Royal. The truth remains elusive but it is probable that the plant riches of Madagascar outweighed for a time the attractions of Paris and the Dombes, and certainly he had less liking for the necessary court-going of a Botanist Royal at the Tuileries or at Fontainebleau than for his work as a botanist out in the field.

Whatever the reason Commerson remained in Madagascar, enjoying the society of the natives whom he described as 'idle and intelligent, mild and yet terrible', making gardens, putting his portfolios in order, and plant hunting to the last. He wrote to the same friend in Paris:

One cannot abstain when in the sight of the rich treasures scattered so freely over this fertile land, from a feeling of pity for those gloomy indoor theorisers who pass their lives in hammering out vain systematics.

The prodigious collector died in 1773 at the early age of forty-six. Eight days afterwards, that is, of course, when they could not have been aware of his death, a full assembly of the Academy of France elected him a member by unanimous acclaim. It was a unique honour because no member had ever before been elected in his absence, and it showed his country's high regard for Commerson's work in the field of botany.

He ranked and still ranks as the most prolifically successful of all plant hunters, making, quite literally, thousands of new discoveries. He claimed to know over 25,000 species of plants. When he died his inventory contained fifty-six huge lists of dried plants collected on his voyage from Rochefort to Madagascar. About 1,500 specimens are now in the possession of the Linnean Society in London, and twice that number are in the Delessert Herbarium. The rest are scattered, but all have been accounted for – except the collection he made on that island of distractions where accommodating ladies danced the timorodee and where ship's carpenter's nails were prized far more than gold. Philibert de Commerson's Tahitian collection was in the first place small, and then altogether lost.

5

Kew and Cathay

Rather less than a year after M. de Commerson, naturalist aboard *La Boudeuse*, had left Tahiti, there arrived in Matavai Bay Sir Joseph Banks, naturalist to Captain Cook's expedition, aboard the *Endeavour*.

Sir Joseph was the Grand Cham of botanical collecting at the end of the Golden Age of Botany. And he was a marvellously lucky man; silver-spooned as it were from his birth in 1743 right to the end of his long and momentous life. Somehow he managed to be at both Eton and Harrow and he had a considerable income from estates in Yorkshire. His health was excellent until he was crippled with gout in later years, and even then he was blessed with a sufficiently equable temperament to put up with his misfortune. He was happily married, had a devoted sister who wrote up his notes, and he treasured the friendship of George III as well as that of humble students, illustrious scientists and the influential statesmen of his day. He was President of the Royal Society for forty-two years, enjoyed immense popularity (except with those who envied his position and constant good fortune) and kept court at his house in Soho Square where the library was cared for by the Swedish botanist Jonas Dryander. Anyone with a curiosity to sell, knowledge to impart, an idea to air, or an interesting tale to tell was sure of a warm welcome – and, if he needed it, a meal and help on his way. But though Banks was a Maecenas, a patron of patrons, he was primarily an active collector, an astute plant hunter whose work on Captain Cook's circumnavigation earned him international honour. To his contemporaries he gave the appearance of being the cockrobin of British Science, but only his enemies refused to acknowledge that he was also the owl. He was that unbeatable combination of a visionary and a practical man, and in a study of plant hunting his chief importance lies less in his own distinguished career as a plant hunter than in his encouragement and training of other collectors to work for Kew.

Kew in the eighteenth century was not as it is now. It was royal not national property;* exposed to the caprice of Hanoverians; to the uncertain taste of Poor Fred's widow (cauterized by Thackeray as a 'shrewd, hard, domineering, narrow-minded' princess); to the benevolent but not very intelligent interest of her son, George III; and, of course, because the family set the pace in energetic breeding, to the influence of their bevy of relations. The Hookers, father and son, were to make it the greatest botanic garden in the world, but only Banks's spadework made this possible. His friendship with George III, his private fortune, and his high prestige both as a practical botanist and as President of the Royal Society enabled him to materialize many of his excellent ideas without a great deal of financial or bureaucratic trouble. It was due to an idea of Banks that Linnaeus's library and herbarium were bought outright by an Englishman and established in London – to the surprise and the chagrin of the Swedes. It was also due to his suggestion that Botany Bay became a convict settlement. Yet the most imaginative of all his plans was surely the establishment of a 'mart and exchange of plants' at the royal garden in Kew.

What is known of Banks's personality suggests that had it been humanly possible he himself would have dearly liked to do all the collecting. But he was a realist, and he had gout. Moreover, he knew that he could come closer to satisfying his scientific appetite if he remained at the centre of things, and just at that particular time he could do no better than stay in London – for London then was England, and England then had a world influence and power she had never had before and would never have again. His decision was to organize from London, to send out his trained collecting commandoes, and arrange the disposition of what their hunting brought in. Most of their plants and seeds found a first home at Kew, but some went to one or other of the botanic gardens in

* At the beginning it was solely a private venture which (according to the *Kew Illustrated Guide*, p. 2) 'explains how it comes about that the most famous botanical garden in the world is situated on a soil so infertile and sandy'.

Jamaica, St Vincent and Ceylon, which were also brain children of the energetic Banks.

The training scheme at Kew was not elaborate and the success of Sir Joseph's collectors all over the world was more a tribute to his assessment of their character and his ability to fire them with enthusiasm than anything else. That he himself had met many of the dangers he expected them to face was in itself an encouragement. Armchair theorizing is one thing. Being warned by an experienced practitioner is another.

Banks was frank. He told his young men that plant hunting required single-mindedness, stamina and a cheerful indifference to discomfort and to continuous disappointment; and, too, that their vocation (he thought of it as nothing less) might involve them in extremes of heat or cold, hunger and thirst, tropical fevers and contact with revolting diseases and creatures, shipwreck and sudden death. The warning came well from a plant hunter who had all the necessary character qualifications himself, and who, in his adventurous collecting, had suffered from exposure in Tierra del Fuego, malaria on the Malay Archipelago, near-shipwreck off Newfoundland, total shipwreck off Australia, and who again and again had had to face the high probability that at any moment his ship would founder on coral reefs, capsize, or be 'pooped' by Pacific breakers. Moreover, as he had noted in his 1769 journal during a storm off the North Island of New Zealand:

The almost certainty of being eat as soon as you come ashore adds not a little to the terrors of shipwreck.

The youngsters would gaze at the fabulous Sir Joseph, admire his sang-froid and candour, and keep him in mind when they encountered dangers in the service of Kew.

Three of them in particular discovered that he had in no way exaggerated, neither overestimating the glory nor understating the difficulty of being a plant hunter; and to this day they are remembered amongst gardeners and plantsmen.

Francis Masson of Aberdeen, trained in plant and seed preservation, actually earned the distinction of being termed 'indefatigable' by the indefatigable Banks. He collected twice in South Africa, and in the Iberian Peninsula, Madeira, the Canaries, the Azores and West Indies, and finally in North America where, it is thought, he froze to death in the icy winter of 1805. He worked for £100 a year with the right to £200-worth of expenses. His triumphs were considerable. Apart from introducing Cape Heaths – which so pleased the gardening public that within fifty years over 400 species were being cultivated – he was also the first to appreciate the value of Cape Pelargoniums, sending home fifty varieties which gave rise to the common Garden Geranium so much loved by the Victorians; and he also introduced varieties of Proteas, Aloes, and the unique Green Ixia, *I. viridiflora*, which on account of its sea-green flowers with the dusky middle caused a minor sensation at Kew.

Against the triumphs must be set Masson's disappointments and hardships. In South Africa there was a wealth of savage animals as well as plants. He was fortunate and mostly escaped their attentions by taking proper precautions, but often, when he trekked into the hinterland to botanize, his life was in jeopardy from hostile Africans. His rule was to penetrate as far as possible into the veldt; that is, until he met armed natives who made it abundantly clear that he could go no farther. On his second expedition to South Africa anti-British Boer settlers were even more menacing than the Africans, and it was not safe to travel more than fifty miles from Cape Town. One especial torment confronted him close to Cape Town itself when a chain-gang of slipped convicts tried to capture him and use him as a hostage. He had an irrational but ineradicable dread of being forcibly held prisoner and he knew quite well that if they had not been chained together they would have caught him. As it was he spent a terrified night, crouched in hiding with a clasp-knife in his hand, while the convicts, growing more and more desperate and threatening, searched for him through the brush in long sweeps. Though on this occasion he escaped, his fear twice materialized in later years: once while he was serving, very unwillingly, as a British conscript at the fall of New Granada in the West Indies and was taken prisoner by the French; and afterwards while sailing to America when he was captured by French privateers. His unenviable death in 1805 brought to an end thirty-three years of sacrifice and strenuous plant hunting in the service of Kew.

David Nelson's career as a Kew Collector was shorter and even more dangerous. Banks sent him off on Captain Cook's third expedition as botanist aboard the *Resolution*. It was an immense voyage which took four years, and in the course of it Nelson was to hunt plants in arctic, temperate and tropical regions. With him on board the *Resolution* was one of Banks's protégés from Tahiti, a youngster called Omai who had been lionized as a curiosity by London Society, painted by Reynolds, taken to the theatre and the House of Lords and (improbably enough) grouse-shooting on the Yorkshire moors. He was now to be returned to Tahiti – complete with a pension from King George, an electrical machine, pots, pans and cutlery, a suit of armour, and a portable organ.

The voyage out was eventful. Having delivered Omai – though not to Tahiti where, as an expatriate, he was no longer welcome, but to a neighbouring island – and having attended a human sacrifice,* when the disembowelling of pigs, the heat of sacrificial fires, the wailing and the drumming, and the piercing shrieks of a little boy which were symbolic of divine gratitude, must have been nerve-fraying even to a Kew-trained plant collector, they sailed north to the Sandwich Islands and there, in a scurry on the beach of sunny Kealakekua Bay, their commander, Captain Cook, was stabbed to death. To the men on the expedition it was a calamity beyond belief. Not only had Cook been much respected, drawing hero-worship from many of them, but undoubtedly he had been the greatest sailor and navigator of his day. Now that they had to do without his genius and his infectious air of confidence, the quite ordinary dangers of the exploration which were bad enough would seem to be multiplied many times. But, perforce, they were obliged to do without him; and they sailed successfully up beyond the Arctic Circle, thence south again towards Japan and China, through the strait between Java and Sumatra, and by the Indian Ocean and the Cape, back home again to England.

So shattering a voyage would have been enough for most young men, but not for David Nelson.

One of Sir Joseph's less successful ideas was to introduce Bread-fruit to the West Indies where he considered it would make cheap and palatable food for Negro slaves. Two expeditions were necessary to accomplish the introduction and prove in the end that he had been over sanguine. Both voyages were commanded by a Captain Bligh, his ship on the first occasion being the *Bounty*. Banks sent his trusted plant collector Nelson, with a Kew trainee, William Brown, to look after the botanical side of things.

The astonishing virtuoso performance of the late Charles Laughton in the film of *Mutiny on the Bounty* has made Bligh into

* John Webber, the *Resolution*'s artist, made a drawing of the ceremony.

an unredeemable villain. It appears from historical account that he did have a sharp manner but that he was no more and no less a brutal disciplinarian than other sea-captains of the day. He has been described as a mariner second only to Cook. Certainly his feat in crossing the Pacific in an open boat and at the same time keeping a tight control over his fellow-passengers was a testament to his seamanship and to his power of command.

Indirectly it was the Breadfruit which caused the famous mutiny. Like a Banana the plant is mainly propagated by shoots from old stock, though it can be grown from seed. Banks's instructions had been explicit. In order to make sure the enterprise did not fail Nelson and Brown were to carry both propagating stock and young seedlings, and this necessitated waiting in Tahiti while seeds were sown and grown into seedlings sufficiently robust to withstand a long sea voyage. Five months in the languorous siren island, where there was very little to do, and each man had his girl, heavily underlined the unattractiveness of life below decks on one of His Britannic Majesty's men-o'-war. The parting from Tahiti was not easy. Some of the crew were genuinely devoted to their *vahines*. Lotus-eating is delicious but to self-discipline it is poisonous. The scupper-gash and bilge-boys who formed a large part of the crew hated the Breadfruit plants as the cause of their leaving the island. Just as much they hated the Captain who was apparently indifferent to the pain of their departure. After a few days at sea, chafing in their cramped and sweltering quarters, and resenting the ration of fresh water lost each day to the Breadfruit, they were ripe for mischief.

David Nelson now had two fresh experiences added to his record of privations: to be abandoned in a small boat in the middle of nowhere; and see the destruction of his plans and hard work – a line of mutinous seamen passing over 1,000 Breadfruit plants from hand to hand to be dropped over the side. His assistant, William Brown, found himself willy-nilly on the side of the mutineers and obliged to join the chain of seamen destroying his beloved Breadfruits. Later he was useful to Fletcher Christian on Pitcairn because of his knowledge of plants and husbandry, but he ended sadly in his own back garden as the victim of a stray bullet.

Nelson's death was of another kind. At the conclusion of that truly terrible forty-seven day voyage of 4,500 miles in an open boat he died in agony of a lung fever.

Bligh recorded in his log:

The loss of this honest man I much lamented; he had with great care and diligence attended to the object for which he was sent.

And, having seen to his decent burying, he noted:

I was sorry I could get no tombstone to place over his remains.

Irascible or not, Bligh was high enough in Banks's regard to be sent out on the second and successful voyage to collect Breadfruits from Tahiti. And, after Bligh had distinguished himself at the battle of Copenhagen under Nelson, it was at Banks's suggestion that he was appointed as Governor of New South Wales to pull together the colony's muddled and slender economy. Bligh's régime there was short, and the quality of his government is still open to dispute, but his continuing part in this study is because there he met and became very friendly with the third of Kew's outstanding collectors, the Yorkshireman George Caley.

As a boy Caley had written to Sir Joseph Banks stating bluntly that he intended to be a botanist. He enclosed with the letter some plants and as one or two turned out to be fresh discoveries Banks replied with a warning that botany brought small financial rewards but if the boy were genuinely keen to learn, a job of weeder at Kew at ten shillings a week was his for the asking. Caley took it, and for two years he learnt as much as he could, then he begged to be sent abroad. Banks at first demurred, then relented, paid for his passage out to Australia and gave him a letter of introduction to Governor Bligh.

For ten years Caley hunted plants in some of the wildest and least known parts of Australia, sending valuable seeds and stock to Kew, and with them loquacious botanical descriptions; on one occasion describing thirty-one packets of seeds in 318 close-written pages. His was a lonely life. After one expedition deep into the interior he told Governor Bligh that in the whole journey the only living creatures he had seen were two crows, adding gloomily that in his opinion even they had lost their way. No doubt because of the monotony he tended to explode whenever he returned to civilization, quarrelling with all his neighbours, threatening once to thrash a missionary, paying overt attentions to a widow and to convict women – one of whom, interestingly enough, was Margaret Catchpole, the smuggler's moll from Suffolk.

Captain Bligh liked Caley; but, then, Bligh was Bligh.

Banks, occasionally severe with errant plant hunters, told Caley to behave himself. But even this Olympian warning made little difference, and so Banks cut off his income, giving him a choice of returning to England or remaining in the Antipodes. Caley chose to return, and took with him an aboriginal manservant named Dan.

Banks had neither the will nor the heart to be stern any longer. He forgave Caley and welcomed him, and because his many plant introductions had been valuable and of great interest, he appointed him superintendent of the St Vincent Botanical Garden in Jamaica. A blunt Yorkshireman – in fact, rugged – to the last, Caley fell out with the Jamaican Quality and closed the gardens on Sundays – the favourite day for visiting – because, he said, they were a light-fingered lot who helped themselves to seeds.

On the death of Sir Joseph Banks in 1820 a Whig Government began to cheesepare at Kew. The garden's allocation was reduced and many of its plant hunters called in.

One of the recalled collectors was Alan Cunningham although he was the King's Botanist; and he was actually refused his claim for compensation for his firearms, tent, provisions and plant-hunting kit which had been stolen by escaped convicts on Norfolk Island. Another, James Bowie, who had served Kew well in South Africa, sending home a quantity of valued plants, including the progenitors of the modern Gladiolus and the first *Clivia nobilis* to be seen in England, found himself unwanted and in financial difficulties. He was obliged to become gardener to a German baron until he was befriended by the remittance man Mr George Rex – a happy-go-lucky bastard of George III, and therefore another of Victoria's 'wicked uncles'.

Kew's prestige fell and was not to rise until Sir William Hooker became first official Director of the National Botanic Garden in 1841.

In these twenty years the chief patrons of plant hunting were a number of recently founded botanical and horticultural societies. They had to pass through the awkward stage and suffer growing pains, some to wither away, others to mature into well-known societies of today. They were responsible for a continuance of the scientific emphasis in plant collection, for encouraging any enterprise which would add to botanical knowledge, and their eyes were instinctively turned to the richest plant repository in the world, when the obduracy of imperial mandarins had made hunting impossible.

At this point it is appropriate to look back in time and examine, although only superficially, the first hesitant and often clumsy plant hunting undertaken by Westerners in mysterious Cathay.

*

In the immense history of Imperial China dynasties have altered, though not often, and imperial pleasure grounds, of course, have undergone some change, but the imperial passion for plants and gardens appears to have remained constant. The first recorded gardening Emperor was the illustrious Chin Ming, who reigned absolute over a large part of the Han Dominions from 2737 to 2697 B.C. The last Manchu Emperor of China, though obliged by political circumstances to call himself Mr Henry Pu Yi, died in 1967, aged sixty-one, and – still true to the tradition of the Dragon Throne – he died classified as a weeder in the Peking Botanic Gardens. There can be no record quite like this in the whole history of civilization.

China gave gardening to Japan, where from the earliest times there was an equally high regard for trees and flowers, for water and for rock formations. But neither country could or would communicate its high culture to the rest of the world.* Largely for political reasons their front doors were kept shut for as long as possible, which made trading difficult and a systematic investigation of their wild floras quite out of the question. Their cultivated plants, grown for centuries in Oriental gardens and smallholdings, gradually found their way to the West and, being unknown and strikingly beautiful, they aroused a furore of interest and, of course, a longing for more. But next to nothing was known of other

* Inevitably, for want of the stimulant of fresh ideas, Oriental gardening became formalized and mannered; it almost atrophied into a fixed tradition. When Westerners were first able to examine the tradition they were enchanted. Chinoiserie in the eighteenth century, and Japonaiserie in the nineteenth, were all the rage.

flowering plants except that there must be large possibilities. The reports on economic plants sent home by missionaries whetted the thirst of Western botanical collectors long before it was legally possible to hunt plants in the Far East.

The Portuguese merchant adventurers took and held Macao. Much later the Dutch tried to do the same with Formosa, but they were driven out by pirates and had to make do with a tenuous footing in Japan. Their base on the small island of Deshima off Nagasaki enabled servants of the Dutch East India Company to report on local horticulture and the island's flora, no more. For the privilege of using the base the fattest of the Dutchmen were obliged to go into Yedo (later called Tokyo) once a year and pay a peppercorn rent of self-degradation. They had to turn somersaults in the street and spit on the Cross. The former humiliation would have hurt their pride but it is surprising that they submitted to the latter – simply for a trading concession.

Spain and Portugal and England, and even China found their trading with Japan strictly supervised. The English East India Company did have a factory on Hirado Island from 1614 to 1623 but it proved highly unprofitable and they withdrew. Their factory at Canton, on the other hand, made a sufficient profit to justify its maintenance but the Chinese were as difficult as the Japanese. The liberty of English sailors and traders was seriously restricted. They were forbidden, for example, to move away from their factory, to employ native servants or to learn the Chinese language; and the supercargoes could only communicate with the government through a selected list of Chinese merchants. The tyranny was endured because in time the trickle of trade became highly lucrative. But it hardly allowed for the collecting and exporting of wild Chinese plants, and it is astonishing that as early as the late seventeenth century a company surgeon called Cunningham was able to collect and dry up to 1,000 species, amongst them seeds of the conifer which bears his name, *Cunninghamia sinensis*. Though officially serving as a surgeon Cunningham was, rather, a negotiator for the company, whose duty was to establish contact with the mandarins and try to open trade with Cochin China. At this fairly high diplomatic level he failed, and so badly that he was the only survivor of a concerted attack on his negotiating party and spent two long years in a Chinese prison. But, afterwards, as a persuader or briber of Chinese guards and coolies, he was an absolute master.

Across the China Seas a German scientist named Engelbert
Kaempfer was enjoying the Dutch Company's hospitality and
carrying out a superficial study of the local flora. His work was
necessarily scanty. He had special privileges and, in fact, saw
enough of the country to write a *History of Japan*, but his liberty
was by no means unlimited. The likelihood of having his ears
pinned to a wooden post and the tender parts of his anatomy
slowly mutilated if he went beyond permitted boundaries must
have deterred his botanic ardour.

A hundred years later the circumstances had barely altered. In
the main Japan rigorously barred foreigners. The Swede, C. P.
Thunberg, who within limits was allowed to collect at the end of
the eighteenth century, and who introduced to Europe the Barberry, *Berberis thunbergii*, was very much the exception. And there
was hardly any improvement in China. Officially by that time some
of the Chinese ports were open to trading but it was not worth the
cost of bribing officials with large presents or permanent pensions.
The Dutch foothold on Deshima, and the ports of Macao and
Canton, continued as the only outlets to the West, and then Canton
itself became threatened. The English government reacted by
sending a plenipotentiary, Lord George Macartney, to the Imperial
Court at Peking.

It was a strange mission: rather overstaffed, certainly ill-
equipped, and altogether improbable in that not one of the prime
officials could speak any of the Chinese languages and they had to
rely entirely on the thirteen-year-old son of the Secretary. But the
mission was cheerful, its optimism founded on that marvellous
English notion that honour and integrity ought, in the end, to
prevail, but that if they didn't there was always the threat of grape-
shot, the cavalry and the flag. As rough-and-ready diplomacy it
proved highly successful in a long history of empire building, but
with a wise and inscrutable sovereign like the ancient Emperor
Kien Lung, who had his own views on honour and was politely
deaf to the claims of commerce, it was a miserable failure.

The restrictions on trading and traders were to continue, and
even tighten, and only one thing prevented the failure from being
total. Sir George Staunton, the Secretary, was an enthusiastic
amateur botanist and the inimitable Banks had, of course, suggested
that two gardeners be added to the British suite. As a result a
certain amount of discreet plant hunting took place on the journey,

and some interesting species were introduced to Europe for the first time.

There was never a large output of cultivated plants through the Dutch trading-posts, but the captains of British Indiamen found it sufficiently profitable to buy plants in the nurseries (called *Fa-tee*) of Canton and Macao, and ship them back to European posts. The 'Eastern Four', that is, varieties of Camellia, Chrysanthemum, Paeonia and Azalea, were in constant demand. Other genera were introduced casually, by chance or by an enterprising seaman. Over ninety 'Chinese' plants were listed in Banks's Kew catalogue of 1810–13, but by far the larger part had been introduced by private travellers or by East India Company servants making a side-shilling from rich gardeners in England. It was unthinkable that Sir Joseph Banks should not poke his finger into this particular pie. No one had a quicker or a clearer eye for a plant and from the ninety in his Kew list he would have made a shrewd estimate of China's possibilities. And so in 1803 he sent William Kerr, a Kew gardener, out to Canton, and if he had not been crippled with gout very probably he would have gone himself.

Banks's vicarious collector did well enough, sending home the first Tiger Lily, which was an immediate favourite, Japanese Honeysuckle, the sulphur-yellow Pigmy Water Lily, the vigorous white climbing Rose which he named *R. banksiae* as a compliment to his master, and another climber which he named *Kerria japonica* after himself. He was based out in the Far East, at £100 a year plus expenses for no less than nine years, botanizing in Java and the Philippine Islands, but spending most of his time in Canton. There he fell foul of the Company's senior surgeon, John Livingstone, a fellow botanist of uncertain temper. Without particularizing, the doctor complained to the baby Horticultural Society of which he was a corresponding member that Kerr had behaved badly in Canton. Moreover, he blamed Kew as well as Kerr because:

> his salary was too small for his necessary wants and consequently he lost respect★ . . . I have not the slightest doubt but his failure is to be attributed, chiefly, to the necessity he was under of associating with inferior persons.

★ This is difficult to follow. £100 a year plus expenses appears to have been the regular Kew salary for a collector out in the field, and the Horticultural Society's own collector in the Far East, Robert Fortune, was engaged for precisely the same figure.

The pompous note excites curiosity; and a longing to know precisely what Kerr got up to, why, how, and with whom. To describe his work as a 'failure' seems overstrong for at the time this peevish letter was written he had been given an appointment at the Colombo Botanic Gardens and evidently had Banks's approval and support. Kerr himself was silent on the matter. He never even explained how he managed to get hold of his plants when the restrictions against foreigners were so stringent. A happy conjecture suggests that he was helped by the 'inferior persons'.

The continuing illiberal attitude of the Chinese mandarins was a source of irritation to the British government. The East India Company complained. Again and again it asked for firm support. Ultimately the government was stung sufficiently to send out a second mission under the leadership of Lord Amherst as ambassador. It promised to succeed better than the first, for the embassy had a Chinese-speaking Secretary; in fact, the very same Staunton who, as a boy of thirteen, had been the only person able to communicate with the Chinese on the Macartney mission. Sir Joseph Banks chose a Dr Abel to travel as the ambassador's naturalist and provided him with two collectors. It was, moreover, a propitious time to send out an embassy because the victory of Waterloo had just made Britain's power plain. Even the Chinese Emperor, it was thought, would have heard of Napoleon and been impressed by his defeat.

But he had not. Or, if he had, military campaigns and victories were as unimportant to him as the claims of commerce.

Again the mission was a failure.

On the day of Lord Amherst's arrival, when protocol demanded that he immediately present his credentials at court, he excused himself on the grounds that he was indisposed. No doubt he was, after the journey up from the coast. But ambassadors have made greater sacrifices for England than attending court while suffering from traveller's dysentery. It counted against Amherst, as did his reaction when informed by an imperial functionary that protocol also demanded he should prostrate himself before the Dragon Throne. He threw a thoroughly British huff and declined to do any such thing.

Making indignant gestures from a position of strength is one thing. From a position of weakness it is another. The Emperor simply declined to see such a proud, pert man and dismissed the embassy. Having achieved precisely nothing Amherst was forced to go home.

On the way back to the coast Abel's two collectors hunted plants. He himself was in a delicate state and he had been a non-functioning naturalist from the start, but he used his influence to persuade the mortified ambassador that, if they slowed down their pace to suit the collectors, the embassy might have at least something to show for all its trouble. Such candour was not calculated to soothe, but Abel was more concerned with botany than Amherst's wounded pride, and, being frank, he got his way. The pace was slowed down and, as they pottered their way to Canton, the collectors foraged alongside the route. They were under considerable pressure, being obliged to keep an eye on their Chinese escort and after a time they had to abandon the general practice of taking duplicates. Nevertheless their collecting was thorough and successful: a rich haul of seeds and plants which, without any doubt, would impress the authorities at home.

But fortune continued to frown on the Amherst embassy.

Dr Abel gave the few duplicates into the care of the embassy's Secretary. The rest was consigned to H.M.S. *Alceste*, and this vessel, driven by a storm, was holed on a coral reef off Borneo, and the main botanical collection went to the bottom of the Java Sea.

In the story of early plant hunting in Cathay, John Reeves holds a distinguished place, not so much because he was an assiduous collector but because he was methodical. He was especially aware of the difficulties in transporting plants over long distances, game to make endless experiments, and never discouraged even when his consignments were lost at sea and he had to make up collections for a second or third time.

In the official ranks of the East India Company one of the most exalted was the post of Chief Inspector of Tea at Canton. Reeves held it for seventeen years and only twice in that long period did he take leave in England. The rest of the time he was wielding a large influence over gardening and botanical affairs both in the Far East and in Europe. In fact, he held a comparable position in China to that of Sir Joseph Banks in England. It was he who entertained visiting botanists, giving hospitality, advice and every encouragement to plant hunters sent out from England.* Inevitably he was Banks's friend and correspondent, and he wrote regularly as well to Secretary Sabine of the Horticultural Society and to Secretary Lindley who followed him. In Canton he made a willing ally of Chief Surgeon Livingstone, and together they devised plans to make the export of plants less haphazard than it had been.

Though a man of such importance, even Reeves was not allowed into the interior and he had to keep within a restricted area round Canton, but being an administrator of great ability he organized illegal squads of coolies to hunt for him in the hills and he paid painters to make a record of the plants that were brought in just in case any should prove to be obstinate travellers and refuse to live through the long journey to England. These coloured drawings were sent to the Horticultural Society in London, actually sold when that Society came to grief, and only found their way back to the refounded and renamed Royal Horticultural Society in 1936.

The monopoly enjoyed by the East India Company in China gave plant hunters the opportunity to collect in the immediate vicinity of Canton and Macao, but one large land frontier to the north was entirely outside the Company's sphere of influence and it was there that Russian botanists made their explorations.

For some inscrutable reason a Russian Imperial Ecclesiastical Mission was housed permanently in Peking and the Chinese authorities permitted the Mission a staff of three scientists. In 1830, Dr Alexander von Bunge, the athletic botanist celebrated for his plant hunting in the desolate heights of Mongolia, was appointed botanist to the Mission in Peking. Unfortunately he had most of the attributes of red-haired men, and no patience at all with the lunatic bureaucracy of the Celestial Empire. Within a year he was causing diplomatic upheavals and his *permis de séjour* was withdrawn. But even in that short space of time von Bunge made fabulous discoveries: amongst them Winter Jasmine, the chicory-blue Leadwort, *Ceratostigma plumbaginoides* (sometimes wrongly called 'Plumbago') and the Spindle Tree which has his name, *Euonymus bungeanus*. A doctor named Kirilof was attached to the same Mission and he was in Peking ten years, but his collections never matched von Bunge's. They included, though, a specimen of *Panax ginseng*, a

* Two gardener-collectors of the Horticultural Society, appositely named Potts and Parks, acknowledged Reeves's kindness in their journals.

plant so rare and so highly valued as an aphrodisiac that the Emperor mounted an expedition every year to hunt for the plants in Manchuria.

The Russians knew something of China's flora, the settlers in Canton and Macao a little more; but the territory actually explored was only a fractional part of the great land mass of China.

Japan, too, remained a fortress, though a Bavarian doctor named Philipp Hans von Siebold had managed to accomplish something. He was attached to the Dutch Station on Deshima and had the inestimable luck to be an eye specialist as well as a botanist. The Japanese, with their noted eye troubles, found it in their interest to allow him special privileges, and he managed to make collections far larger than his predecessors in Japan, Kaempfer and Thunberg. His life, however, was not without risks. Bartering his ophthalmic skill for a licence to move freely on the main islands, he had seduced the myopic Imperial Astronomer into letting him have a copy of a map of Japan, a secret document of great importance, but on his way back to Deshima he was shipwrecked and the map discovered. Japanese heads rolled – but an eye doctor was an eye doctor. Von Siebold was sentenced to two years' confinement to Deshima, to be followed by banishment. In 1830 he took 458 plants with him to Holland; in itself a considerable achievement. But he had seen the possibilities on his trip to the main islands. He described them. Botanical appetites in Europe were stimulated.

At the beginning of the nineteenth century the forbidden interiors of both China and Japan held golden promise and the restrictions must have been agonizing to plant hunters of imagination and spirit.

Away to the west in North America there were no restrictions beyond those imposed by wars, wild Indians, deplorable conditions and dangerous terrain. Because of their preoccupation with settling the land and spreading frontiers, establishing agriculture and commerce, and fighting Indians, Englishmen and the French, few American-born colonists had had time for plant hunting. John Bartram was an exception, and so were Lewis and Clark and Long, all of whom collected on their expeditions to the west. Otherwise the scientific botanizing in the States had been undertaken by Europeans, by André Michaux who collected for the French government, by an Austrian, Frederick Pursh, by two Englishmen,

Nutall and Goldie, and by two Scots, Archibald Menzies and David Douglas. The last made his collections when Kew was at its lowest ebb and the Horticultural Society of London tumbling into ruin, and, without any doubt, he was the most intrepid and successful hunter ever known.

6

David Douglas of Scone

It probably has no significance but is none the less remarkable that, whereas emperors, holy popes, kings, electors, reigning dukes and bishops have generally been enthroned on worthwhile articles of furniture, the Kings of Scots have not. Their throne has been stone; simply a rock, greyish-blue in colour, cold to the touch, roughly the size and shape of a church hassock. Scone, where the coronations took place, once had a substantial royal palace. After the Union it fell into desuetude. What was a royal city, a meeting place of parliaments, shrank to finish as a decayed, run-down hamlet with a handful of cottages and a church.

It was there, right at the end of the eighteenth century that David Douglas was born.

His father called himself a stonemason, which meant that he turned out a tombstone or two and cut lettering, but otherwise was the district's jack-of-all-trades. He was noted for being short-tempered, unsmiling and mulish.

All three handicaps were inherited by his second boy David. The resulting clash of wills was so alarming that Mrs Douglas got the boy out of the way by sending him to the village school at the tender age of three. This, at any rate, is what she hoped; but David defied the schoolmistress's authority, played truant, and behaved so badly that he had to be taken away. Either that or he was expelled – and all before the age of seven.

At his new school the master believed in using the tawse – a double-ended strap of hardened leather – very hard and very regularly, but it made small difference to young Douglas. He was no more respectful or punctual than before, neither was he any more thorough with his schoolwork. Probably the only thing which prevented this bumptious delinquent from developing into a thug was his consuming passion for natural history. It humanized him, and good luck helped as well for when, to everyone's relief, he left school at the age of ten, just the sort of job he wanted was waiting for him on his own doorstep.

No large garden was within reach of Scone until about the time David was sacked from the village school when the Old Palace was rebuilt by the Earl of Mansfield. Considering Scone was too close to the palace he had the hamlet moved, literally, two miles farther away. The parish church was taken down, its stones numbered, and rebuilt on the new site. Over £70,000 were spent on the mansion, which had 125 rooms, ninety of them bedrooms, and this magnificence was matched in the grounds. The gardens, already becoming well-known for their excellence when David Douglas left school, were superintended by a friend of his father, the head gardener, William Beattie.

Scone Gardens and Mr Beattie worked like an Aladdin's lamp on the boy. His pigheadedness and his idleness in school changed overnight to a willingness and eagerness to learn. In his seven years' apprenticeship he became a master in the practical details of working ornamental, forcing and kitchen gardens. He also became proficient as a botanist, collecting wild specimens in the hills for drying and pressing, or transplanting to his father's garden. He became thirsty for knowledge; reading travel books, works on natural history, the Bible, anything he could lay his hands on; and when, at eighteen, he was recommended by Mr Beattie to a post in the famous gardens at Valleyfield near Dunfermline, he seemed well set on the path to distinction as a master-gardener himself. But the large collection of exotic plants at Valleyfield suggested a fresh and very original idea to Douglas. Instead of a gardener, he would be a botanist, and not an ordinary, stay-at-home, microscope and notebook botanist, but a collector out in the field, a real hunter.

It has already been remarked that the rise to fame and fortune of botanists from humble circumstances reads very often like a fairytale. David Douglas's good fortune was no less exceptional.

To begin with his employer at Valleyfield gave him the privilege of using his own private botanical library and he took such advantage of the opportunity that in two years he was able to apply for, and get, a place on the staff of the Botanical Garden at Glasgow. There he caught the eye of William Hooker, one of the foremost botanists of the day, and Professor at the university. The country boy from Scone and the Professor of Botany became fast friends, and botanized together in the Scottish Highlands and Islands. So high was Hooker's regard for his young companion that he recommended him to Secretary Sabine of the new Horticultural Society of London, and at the slender age of twenty-four Douglas was given his heart's desire by being appointed to collect information and plants for the Society in the east of the United States.

*

A map of the United States in 1823 looks quaint.

Parts of the country were unexplored, huge areas unsettled, even unmapped. The horn of Florida had recently been bought from Spain. The present states of California, Nevada, Utah, Arizona, New Mexico, Texas and three-quarters of Colorado were colonies of the Mexican Emperor. Washington State, Oregon and Idaho were still English, and Alaska was Russian. Though New York boasted a population of 125,000, there was a neglected twenty-acre garden where the Rockefeller Center now stands, and the city scavenging department was composed of a herd of gutter-rooting and very healthy pigs.

Douglas's job was to find new plants if he could, but principally he was to study American fruit-growing in the east and beg or buy specimens of trees.

The voyage out from Liverpool took fifty-nine whole days. On the second week fresh water was rationed, and tobacco became so scarce that the crew used it twice over – chewing it first, then drying it in the sun to make it smokable. Getting ashore at New York must have been like escaping from prison.

Douglas's first expeditions at once illustrated his determination and singlemindedness. He met some colourful characters, amongst them William Coxe, an orchidist who gave him for presentation to the Horticultural Society two bottles of seven-year-old cider, and an ex-governor of New York, DeWitt Clinton, who consigned to his care six wild pigeons, boxes of minerals, apples, pears, moccasins, books and written speeches. He travelled by various conveyances – stage-coach, horse-drawn canal barges and paddle-steamers. In one of the latter they hit a storm and a paddle was swept away. He saw some fine sights, too; the house of exile of Joseph Bonaparte, Napoleon's brother and ex-king of Spain, the Niagara Falls, and the gala opening of the Erie Canal at Albany. But he thought of little except natural history. The botanical gardens at Philadelphia and Bartram's Garden of Delight at Kingsessing mattered a thousand times more to him than the pleasure grounds of Bony's brother. The best part of his trip to Niagara was the opportunity to botanize on the river banks and on Goat Island; and while Albany had a rip-roaring time at the canal-opening with the firing of guns, bands, processions, fireworks, booze, barbecues and general uproar, Douglas was locked in a hotel room arranging his collections.

His is one of the most extreme examples of a boy's hobby turning into a man's full-time job, of an interest becoming an all-absorbing passion – what, quite rightly, is called a craze. Very probably the citizens of Albany regarded their Scottish guest as crazy.

Douglas's trip, anyway, was successful. He carried back to England many plants of interest, notes on their distribution and habitat, seeds, dried specimens and living plants; and, in addition to the gifts from Coxe and Clinton and other well-wishers, he took with him some living wood ducks and quail.

The voyage home was shorter and more comfortable: barely thirty days. Curiously enough, only the ducks were seasick.

*

Douglas's other trips for the Horticultural Society of London could be called his Great Adventures. In them he often trod where no white man, and perhaps in some cases where no one at all, had ever been before. They were in the wilds of Canada under the protection, so-called, of the Hudson's Bay Company or on the undeveloped, mostly unexplored western coastline of North America.

The easy way to the west coast was round Cape Horn and on his first long voyage out aboard the *William and Ann* in 1824 Douglas was delighted to have the company of a youngster he had known at Glasgow University: John Scouler, aged nineteen but already an M.D. and ship's surgeon.

Scouler was as much a magpie collector as Douglas. At Rio, for example, we hear of him returning on board one evening looking extremely odd. The japanned collecting tin on his back was jammed tight with plants, so much that they spilled out of the top.

His pockets were full of pieces of granite. His hat inside and out was pinned all over with insects. Both arms were full of plants, and somehow he also managed to be carrying one of the most venomous of Brazilian snakes.

He and Douglas packed two boxes of rare orchids and other plants and sent them back to London from Rio, and on the voyage south to the Horn they continued their observation of natural history: measuring seaweed sixty feet long, and catching several albatrosses, one with a twelve-foot wingspan. What the superstitious sailors thought of this is not recorded.

It was a happy voyage for both of them. The captain being tolerant (he was rewarded for his patience by a silver medal from the Horticultural Society) they turned his ship into a floating laboratory. When they made landfall at Fort George (now called Astoria), and while the *William and Ann* was being refitted and re-victualled, Surgeon Scouler was given permission to botanize with Douglas. Apart from the company of a few enthusiastic amateurs amongst the 'gentlemen' of the Hudson's Bay Company and the 'employees' who were mostly French Canadian, Indians, half-breeds with a few Sandwich Islanders, or friendly Indians whom he trained to collect for him, Douglas thereafter was a lone plant hunter.

On New Year's Day, 1826, after six months in the Columbia River Region, and having already suffered a great deal for the sake of botany, Douglas wrote in his journal:

> I am now here, and God only knows where I may be next. In all probability, if a change does not take place, I will shortly be consigned to the tomb.

Of Douglas's three main expeditions, the first to the east coast, the second in the north-west and across British North America and a third in California and the west coast, the second made the greatest demands upon him.

He proved his devotion, or craze, by once undertaking an extended journey into unknown territory provided with hardly any clothing at all, simply so that he could carry thirty quires of paper, weighing 102 pounds, for drying and packing the plants he hoped to collect. On another occasion when snow-shoeing through the Rockies with a party of hungry, half-frozen Company men (an appalling journey on which they had to ford a racing mountain river fourteen times in a day by plunging up to the waist in the bitterly cold water), one of them was lucky enough to see and shoot a game bird renowned as excellent eating. But no one had a taste of it. Douglas was there first, and not having seen such a fowl before he refused to give it up, stuffed it on the spot, and added it to the heavy load he was already carrying. He named it Franklin's Grouse and the very specimen which he roughly preserved about 150 years ago in the Rockies may still be seen in the Royal Scottish Museum in Edinburgh.

Apart from the sheer fatigue of travelling huge distances over uneven ground on foot, he faced some very real dangers. He made friends with a few of the Indians, especially amongst the Chinooks, who called him 'The Grass Man' for obvious reasons, and who considered he must be one of King George's great white chiefs. These were friendly, and from others he bought friendship with beads, buttons, rings, gimcrack trinkets and plugs of tobacco, but quite frequently he found himself threatened by hostile Indians. There was always the risk of meeting a war party or employing a treacherous guide.

He had to put up with worse trials: long bouts of hunger so extreme that he was next-to-starved and reduced to eating all the berries and seeds he had been at such pains to collect, the roots of wild Arrowhead and Liquorice, the skins of the animals he had dried to send to England, and 'twice being obliged to eat up his horse'. Then there were natural dangers such as slime-pits, snow-drifts, chasms, avalanches, half-hidden rocks and other river snags that tore the bottom out of a canoe, rapids, waterfalls and floods. He lost all his collections twice in the same hazardous place. There were fevers as well; which only the toughest of the Europeans could withstand and which decimated the Indians. Attacks of fever weakened Douglas, as did a festering knee caused by a rusty nail, and he began to suffer from a severe irritation in his eyes which made seeing more and more difficult. Vermin were an additional menace. While he lay asleep, pack rats ate up all the seeds he had recently collected, gnawed through a bundle of dried plants, carried off his razor and shaving bowl, and did not wake him until they began to drag away his portable inkstand. This disturbed him. He stretched out for his gun and shot them dead. He also had a bad experience with 'an indescribable herd of fleas' in an Indian hall. He escaped from them to sleep outside, where immediately he was attacked by 'two species of ants, one very large, black, three-quarters of an inch long, and a small red one'.

Poor weather was so commonplace that for days and nights on end Douglas was never dry, and the storms could be wild and frightening:

> The rain, driven by the violence of the wind, rendered it impossible for me to keep any fire, and to add misery to my affliction my tent was blown down at midnight . . . Every ten or fifteen minutes immense trees falling produced a crash as if the earth was cleaving asunder, which with the thunder peal on peal before the echo of the former died away, and the lightning in zigzag and forked flashes, had on my mind a sensation more than I can ever give vent to.

His fortitude was nothing short of heroic and it is barely surprising that his already short temper grew shorter. Candidly he admitted it:

> . . . travelled thirty-three miles, drenched and bleached with rain and sleet, chilled with a piercing north wind; and then to finish the day experienced the cooling, comfortless consolation of lying down wet without supper or fire. On such occasions I am very liable to become fretful.

Most of us would weep, turn up our toes, and die.

It was on this, his second American expedition, that Douglas found and introduced to Europe the Pine which bears his name: a massive but graceful tree, and so tall that, in its native forests, the branchless trunks soar upwards like cathedral pillars of green-grey marble. And the Douglas Fir was only one of his many finds amongst the conifers. In a letter to Hooker in Glasgow he wrote: 'You will begin to think that I manufacture Pines at my pleasure.'

Perhaps the Sugar Pine was the most beautiful of them all. Certainly it led him the greatest dance.

Right at the beginning of his exploration he had found Calapooie Indians chewing large sweet-flavoured pine seeds, and learnt that they came from an outsize tree which grew far away. A little time after, having established that the trees grew in the mostly unknown country of the Umptqua Indians, and learning that Company fur hunters were about to explore that region, he attached himself to the party.

The fur hunters rode but the plant hunter preferred to walk and use his horse as a baggage carrier. They had to rely for food on what they shot and game being in short supply, quite soon they were hungry. This was not the least of their trials. One of the party was treed and partly savaged by a grizzly. By the time they had reached the site of the present city of Elkton, the fur hunters were exhausted through hunger and from hacking a trail through thick undergrowth, and their horses were worn and winded.

The indomitable Douglas pulled in his belt another notch, went out botanizing on his own, toppled into a deep gully, and lay unconscious for five hours until some Calapooies found him. His chest was very painful, so he gave himself a drastic kill-or-cure treatment, bled himself from the left foot, and took a bathe in the ice-cold river.

Against all the rules, he recovered.

At this point he decided to leave the main party and set out after the Sugar Pines guided by the son of a chief called Centrenose.

It was a terrible journey.

They had enough food for one meal a day, were drenched and frozen in a storm, and the botanist caught some sort of chill. His head was racked with pain, his belly tightened with cramps and he had such dizzy spells that he could barely see. He cured the chill by sweating it out and then tramping thirteen miles.

Eventually, when walking by himself, he met an Indian who strung his bow at the ragged collector and would have shot him. Douglas quickly laid down his own gun, gave the boy some beads, shared a pipe with him, and by sign language discovered that he was about twenty miles from the Sugar Pine country.

He reached his objective some hours later and forgot his tiredness and hunger and was transported with delight.

The trees were too vast to be measured as they stood, but one, which had been blown down by the wind, had a total length of 215 ft, and at its fattest was 57 ft 9 ins round. Douglas carefully noted these details and others, and then set about shooting down a few of the foot-long cones from the tops of the trees.

There he was, blasting away, chipping them from the branches with ball, when he found himself surrounded by a party of eight

Indians in full warpaint who were armed to the teeth and looked anything but friendly.

A battle of nerves followed: the Indians with strung bows aiming at Douglas, and Douglas six paces away with a cocked gun in one hand and a pistol in the other. It went on for eight, nine, ten tense silent minutes, until the Indian leader abruptly asked for tobacco.

The tension was broken. Douglas promised them tobacco if they fetched him more of the cones. They agreed and disappeared.

He picked up three bright cones he had shot from the tree and ran as fast as he could back to his camp. There, unable to converse with his guide, not daring to light a fire, and supperless, he spent an uneasy night in constant expectation of an attack.

No attack came. Instead, three grizzly bears turned up at dawn. Douglas shot two of them.

He then made his way downriver; and, meeting the main party, they built fires against the possibility of Indian attacks, pulled in their belts – for they were all famished with hunger – and, the strain getting the better of them, quarrelled fiercely amongst themselves.

Douglas's worst experience came near the end of the journey, when, in swimming a river, he lost by far the greater part of all he had collected in the past weeks of continual hardship. Happily, though, he still had his precious Sugar Pine cones.

After such a harrowing adventure anyone less than David Douglas would have taken a rest, particularly as he suffered afterwards from severely swollen ankles caused by walking so far in the wet and cold. But once his feet were in trim again he could not sit still. He was off to the coast by himself to collect seaweeds and shells – and there a tempest dashed his canoe to pieces on the shore. He crawled to safety and took refuge from the storm with an Indian friend who fed him on dried salmon and Salal berries. This unsuitable food gave him such severe and weakening bilious attacks that his long trek back to the Company's base was a torment of exhaustion and hunger. He arrived there on a Christmas Day, rather more dead than alive.

Douglas had his rewards. His successes were enormous.

After this expedition he made the difficult journey overland to take ship on Hudson Bay. When he reached York Factory on the south-west shore he had travelled by canoe, or shallop, or on horseback, snowshoe or on foot over 10,000 miles in a little over two years. Without a companion or crampons, alpenstock, snow goggles, or even a walking-stick (and simply as a diversion in the middle of a long trek) he had climbed the two highest peaks in the Rocky Mountains and named them in honour of Brown the botanist and his great friend William Hooker. He had set a record in finding and ultimately introducing to England more plants than ever before had been introduced by any individual from any country.

Fate played a last little trick on him a few days before he was due to sail on the *Prince of Wales*. Because of shallow water the ship lay out at some distance from the shore and Douglas, with a party of three and eight oarsmen, rowed out to visit her riding at anchor. On their way back they were caught in a hurricane and swept out into the bay. The waves were many feet high. There was tremendous thunder and rain. The cold was intense. Rowing was impossible. All they could do was try to keep the boat afloat by bailing continuously with their hats. This they did all night, all the next day, and well into the second night before they could use the oars. By this time they were seventy miles out without a compass and in the pitch dark. It was a miracle, nothing less, that eventually they got back to York Factory; all the same they suffered severely from the cold and the exposure. Douglas especially was in poor shape, and he hardly recovered the use of his limbs before the *Prince of Wales* made landfall at Portsmouth on 11 October 1827.

David Douglas had been out of the country for a little over three years. He was twenty-eight years old and already famous.

*

The magnificent achievement of their salaried collector was genuinely appreciated by the Council and Fellows of the Horticultural Society of London, and, too, by a far wider circle of admirers. Douglas was elected a Fellow – and without the obligation to pay fees or subscriptions – of three learned bodies, the Linnean Society, the Zoological Society, and the Geological Society. Arrangements were made for his journal to be published.

He was gratified by the honours and, to begin with, he enjoyed being made much of in London drawing-rooms and committee rooms, even ballrooms. Either it then went to his head or (and this is more probable) he became bored. Certainly at what could have been one of the happiest times of his life, he was discontented, surly, anxious to find fault.

He had some cause. To have made such sacrifices for the Horticultural Society – and then discover he was being paid less than the porter who opened and shut their front door would have inflamed a less excitable man than David Douglas. And his bad temper was nourished when he discovered that some of the natural history specimens that he had so carefully sent home in 1825 had been thoughtlessly stuffed in a corner. They were moth-eaten, crushed, and useless.

He quarrelled with Secretary Sabine. He quarrelled with the Committee members. He quarrelled with the Fellows. He would have quarrelled by post with Hooker if Hooker had allowed him to. Hooker more than anyone knew what was wrong, that restlessness was at the root of the trouble, and being lionized and made much of did not suit David Douglas of Scone.

The Scots, like Slavs and most European Russians, are noted for their taciturnity. If it is relieved by a sense of humour, however wry, all is well. Douglas was simply taciturn, as unsmiling as his father.

The Scots, like Texans, Arabs and the hill tribesmen of India, are fiercely independent people – which is honourable and praiseworthy if it is not spoilt by pigheadedness. Douglas's early history showed how large a share he had of stubbornness.

He was a misfit in London society; dour, humourless, suspicious and cantankerous, and in the end ill-mannered. Had he had someone to love he might have been different, but nothing in the nature of a love letter has ever come to light in all his private correspondence. He respected and admired a few friends. He was devoted to a small Scots terrier called Billy which from this time was hardly ever away from his side. But he only gave his heart to plants.

None of this says that by nature he was an unhappy man. Dickens probably made a mistake in supposing that old Ebenezer Scrooge was miserable simply because he was unlike everyone else in *A Christmas Carol*. In his own grouchy, Scrooge-like way no doubt he was quite content without human companionship, goose and plum pudding. And Douglas may have been the same. In

London society he behaved badly because he was as much out of his proper element as a fish in a tree or a bird in a goldfish bowl.

Hooker knew this. Writing many years later he described Douglas's low spirits and irritability and the cause of both:

His temper became more sensitive than ever, and himself rest-less and dissatisfied; so that his best friends could not but wish as he himself did, that he were again occupied in the honourable task of exploring north-west America.

There was widespread relief when the Horticultural Society sent him off on another expedition to the western coast of North America. He was to travel by way of the Horn and the Sandwich Islands and concentrate most of his attention on the flora of California. The Zoological Society sponsored him in a small way, as did the British Colonial Office who wanted accurate maps of the Columbia territory. He was shown how to fix geographical positions with their longitude and latitude, given a sextant, chronometer, barometer, thermometer, hygrometer, compass, the most expensive apparatus for making magnetic observations, and taught how to use them.

Douglas was being sent back into his proper element under the patronage of the British government, the Royal Navy, the Hudson's Bay Company, the Horticultural and the Zoological Societies. As he noted himself, it was 'not the journey of a commonplace tourist'.

He sailed away on 31 October 1829, and never saw England again.

*

His previous hardships had told on Douglas. When he reached the Columbia Region after an eight-month voyage, his old acquaint-ances noted the alteration in him. During his absence the Green-wich Naval School had sent out several youngsters to be trained by the Company. One of these boys, who had never seen the thirty-year-old botanist before, described him as 'a fair florid partially bald-headed Scotsman of medium stature gentlemanly address about forty-eight years of age'.

He was still tough, though, under his battered surface. He hired a man-o'-war's deckhand to carry his equipment when he went col-lecting, and for the time being his journeys were neither as long nor as adventurous as before, but he had the resistance to survive a bout of fever so terrible that whole Indian villages, of 450 inhabitants and more, were totally wiped out, twenty-four of the Company's men died, and trading for a while was at a standstill. Douglas apparently kept himself on the go with what he called 'healthful perambulations', in other words, arduous botanizing trips.

By the time he judged he ought to be moving south to California, he had collected a consignment of dead and living quadrupeds, birds, reptiles and insects for the Zoological Society, and, for the Horticultural Society, three chests of seeds, and over a hundred new species of plants – including six unknown Pines.

As the Indians down in the Sugar Pine country between Columbia and California were restless and dangerous, and Douglas felt less inclined to take risks than he had before, he decided to travel by ship.

Archibald Menzies, naturalist aboard the *Discoverer* on Vancouver's exploration of the western coast of America, had landed in California even before Douglas's birth. He had collected a few plants and written careful and tantalizing descriptions of others. At the time of Douglas's visit it was a Mexican colony where the vine, the olive and the fig, and even bananas and sugar-cane, were cultivated; but the larger part was unpopulated and generally unknown.

Douglas showed the uncertainty in his mind of what might lie ahead of him by the list of provisions he bought before leaving Columbia. Apart from the obvious necessities like shoes, shot, codline and candles, he included ten pounds of tea, nine gallons of brandy, two large black silk handkerchiefs, a large moose-skin, and a Jew's harp. He paid off his deckhand servant, and with Billy the terrier took a brig sailing south.

'Don David' as Douglas was called because he was an '*hombre de educacion*', was enchanted by California: its rich flora, the kindness of the inhabitants, the hospitality and help given him by the Franciscan friars who 'loved the sciences too well to think it curious to see one go so far in quest of grass'.

The only drawbacks to this otherwise perfect place were the presence of two other botanists* and the intense summer heat. He

* Dr Thomas Coulter was collecting there for a Swiss patron, and Ferdinand Deppe, a German naturalist who travelled the world as a supercargo.

made friends with the former and they swopped finds like school-boys, but he could do nothing about the latter and it was 'hot enough to suffocate all fleas'.

The spring was unbelievably beautiful and it was then that he had to work speedily before the heat parched up the vegetation. In his first year, Douglas arranged and catalogued a number of unknown genera and over eight hundred species, many of them new.

In California he found and later introduced to London the Garrya which in its male form is one of the most striking and elegant evergreens under the sky. Its egg-shaped leaves are the colour of emeralds above, and underneath are furry and grey. The catkins – a foot long in the States, but barely half that in England – are like closely linked necklaces of jade. This, with the introduction of the Monterey Pine and a number of brilliant annual flowers like Wild Heliotrope, Blazing-star and Californian Bluebell, gave him fame far beyond Great Britain. For example, a duplicate set of one of his collections was sent to the Botanic Garden at St Petersburg.

The especial value of Don David's Californian expedition was that he stayed nineteen months there and was systematic in his collecting. In the end he had about sixty new plants, but his collections consisted of hundreds of species, besides seaweeds, mosses and liverworts.

In the soil attached to their roots were flakes and spots of a yellowish mineral. Douglas knew what it was but he did not guess how much of this yellow mineral there was. If he had, perhaps the great Californian Gold Rush of 1848 would have taken place seventeen years beforehand.

Douglas returned to Columbia, via the Sandwich Islands, for another collecting trip, and heard there that the Horticultural Society which employed him was in a turmoil. His patron Sabine was such an enthusiastic gardener that he had handed the everyday affair of the secretaryship to a clerk – who had embezzled the Society funds. At the same time it was discovered that certain Fellows owed outstanding subscriptions, including the King, a duke, a marquess, thirteen other peers, and twelve clergymen. Sabine resigned. Douglas out of loyalty resigned with him. Thereafter all his collections went to Hooker.

But this was not to be for long.

A project approved and backed by the Czar and the Russian Governor of Alaska, that Douglas should botanize across Alaska, Siberia, Russia and Europe ended only 1,150 miles after the start of the journey.

Even covering that distance was an heroic feat because Douglas's eyesight had worsened. He was completely blind in the right eye and had to wear smoked spectacles to protect the other. As a result his temper had shortened and sharpened. He had quarrelled with a Company trader on the journey and had been challenged to a duel – which he declined. He learned that the Indians ahead were unfriendly. He was low in spirits and could not face the prospect of months of travel through unknown country. For the first time he turned back. A week later came disaster.

His party was canoeing down to headquarters on the Frazer River and, in the rapids at Fort George Canyon, the canoe was ripped and smashed to pieces. Douglas himself was carried down to a whirlpool and swirled round and round in dizzy turbulence for an hour and forty minutes before being washed out like a cork on to a rocky shore. Providentially all the travellers escaped with their lives, but Douglas had lost a collection of 400 species of plants and – this was the real disaster – his volume of field notes in which he had listed, against numbers, the precise location of all his important Californian plants. It was a serious loss to him, and has been to botanists ever since.

He reached the Columbia River exhausted. Three months afterwards, and with enormous reluctance he decided he would have to return home. To do so by the Sandwich Islands, where he could collect from the tropical vegetation and examine the volcanoes, made the thought of London fractionally more bearable.

He and Billy sailed in the Company's brig in October, 1833 – touching for a time at San Francisco – 'a few dilapidated buildings' – where Douglas camped and botanized on an eminence which is now roughly the centre of modern San Francisco. Thence he sailed to Honolulu.

Douglas enjoyed his months in the Sandwich Islands. He was intrigued by the natives' notion that fatness is attractive, and noted how they stuffed themselves with *poi* – a starchy dish made from Elephant's Ears. He hired porters, a cook, a bird-catcher and a guide who carried his spy-glass and umbrella; shot blacks pigs and wild cattle (the descendants of those brought to the islands by Captain Vancouver); botanized, finding some interesting ferns as well as many mosses and lichens; and travelled the islands in turn making magnetic and geographical records and observing extinct and half-active volcanoes.

By the end of June he was familiar with all the islands and felt obliged to face the depressing prospect that, when a ship bound for England made landfall at Honolulu, he would have to leave for home. While waiting, he undertook to show a visiting missionary the wonders of Hawaii. They agreed to meet at a village on Hawaii called Hilo.

The meeting never took place.

Douglas walked the long route over the mountain. He was unable to hire a porter and carried his own small bundle and equipment. Billy trotted with him. He stayed the night at the house of a man named Davis, and, early the next day called on an English ex-convict, Ned Gurney. He was warned that about two miles farther on there were three cattle traps, that is, crudely-made pits disguised with branches, leaves and earth.

Some time after, two of Gurney's natives heard the unmistakable sounds of a trapped beast in one of the pits. They looked in. Snorting with fear, and stamping in the debris of soil and leaves which had caved in under its weight, was a terrified wild bull. Beneath it, half-covered with fallen earth, the clothes rent, lay the gored body of a man.

In the pathway, a little black dog was guarding a bundle which the natives recognized. The dog howled and howled and would not stop.

No one could be sure whether Douglas had met his death by accident or not. Old lags are never free of suspicion. It was said that the botanist had had a bag of money on him when he set out, and that this was never found. An American cattle-hunter examined his body and expressed the view that the wounds and gashes could not have been caused by a bull. But there was sufficient doubt for the local missionary to embalm his body so that it could be sent to Honolulu for expert examination. There the experts said the wound had been caused by a bull.

In his book on Douglas, A. G. Harvey* has made the point that there are unconvinced people who believe he was murdered, but it seems probable that Douglas heard the bull's bellows and, like the natives after him, peered into the pit – and slipped. His end, with no chance at all of escaping that maddened beast, must have been a terrible one.

He was mourned by men of science all over the world. But he only left £50 and a bundle of dried plants collected in the Sandwich Islands.

Billy was sent back to England.

* A. G. Harvey, *Douglas of the Fir* (Cambridge (Mass.), 1947).

PART TWO

The Techniques of Collecting

i

The Carriage of Plants

Transporting Exotics: a Short History of the Carriage of Plants: Particular Difficulties on Sea Voyages: Experiments, Recommendations and Complaints: Transporting Seeds

ii

Nathaniel Ward of Wellclose Square

A Portrait from Deduction: the Caterpillar Jar: Ward's Observation and Reasoning: the Prototype Design: *Terraria* – their Decoration and Contents: the Fern Age: 1834, the *Annus Mirabilis* because Ward's Thorough Experiment Succeeded: Sages and the Eminent Acknowledge their Debt to Ward: the Use of Wardian Cases in Introducing Economic Plants: a Dramatic Change in the Prospects of Plant Hunters and their Patrons

The Carriage of Plants

As the world becomes increasingly materialistic the international posts become, rather surprisingly, more costly and eccentric; but it is still possible to send by airmail a plant in an airtight and rigid plastic container and – trusting that it meets with neither an excise probe or post office neglect – have a reasonable hope it will arrive safe and sound and in time on the other side of the world. In the early days of plant hunting, in fact right up to the nineteenth century, even to have dreamed of such a Cloud Cuckoo Land of transport would have invited the attention of the Masters of Lunacy.

Queen Hatshepsut's hunters had the right idea in 1482 B.C. when they transported trees and plants in reed baskets and jars of soil, but they merely had to carry them through favourable climatic conditions from the Land of Punt to Egypt, and, moreover, they had at their disposal Pharaonic slave gangs to carry fresh water, work fans, adjust sunblinds and do anything else necessary to the health of their green treasure.

Not a great deal of progress was made in the next 3,316 years. True, David Douglas had rolls of oilcloth in which to pack his dried plants, waxed containers for his sleeping bulbs, corms and rhizomes, and little seed packages treated (ineffectually) against vermin and bugs, but he was alone and without the help of helots, and the ship which carried his collections on the long journey from Vancouver Island to London hardly made more speed than Egyptian vessels plugging up and down the Nile. Douglas sent living stocks as well, but, like Gobi-crossers, only a tough or lucky few got through. In fact, the chance of keeping plants alive on long journeys when communications were, to say the least, sluggish, was about equal to the chances of messages in bottles and those attached to gas balloons arriving safely at their proper destination.

Even hunters on elaborate expeditions who had the advantage of coolies to carry boxes of plants found that over large distances it was difficult to guard against killing the weakest plants outright, and making valetudinarians of the strongest by subjecting them to extreme variations in temperature, humidity and altitude. South America has always bred extremes in almost everything – not least in the difficulties a plant hunter encountered in transporting his finds to the coast. The journey would be dangerous, long and tedious, often lasting many weeks. It would involve waiting for fresh mules or porters, for canoes and native rafts and steamers, for transport of all kinds at different points along the route, sometimes for tracks to be cleared of fallen rock, bridges rebuilt, for floods to subside and dried-up rivers to refill. At regular intervals the boxes would need to be opened and their contents inspected, the diseased and damaged plants jettisoned, the rest repacked. The plants would feel the change from climate to climate, of varying temperatures, humidity and light intensity, as they were carried through swamps, grass plains, difficult jungle country, primaeval forests, over high mountain passes and on the breast of wide rivers either stinking in tropical heats or made turgid by the rains. Orchid hunters in particular discovered that in South America it was not sufficient to have a nose for a plant and the courage and stamina to follow its lead. They also had to be expert in transport problems and plant management, capable of dealing with muleteers, overseers, rivermen and supercargoes.

At the port the hunter's responsibilities ended. His anxieties did not. For all he knew the captain would redistribute the cargo out at sea; have the plants moved forward where the danger from sea-spray was a hundred times greater, or stowed below where the conditions were exactly right to ferment the plants into putrid scraps of vegetable matter. It would be done 'in the interests of good seamanship', but, in reality, for the captain's own personal comfort. Not many enjoyed having their poop and their living-quarters cluttered up with plants.

Even supposing the captain was a plantsman himself, or that he had a particular personal commercial interest in the consignment

1 A carved relief on the temple walls at Karnak showing trees in pots collected on Queen Hatshepsut's expedition to the Land of Punt in *c.* 1482 B.C.

J.ᵗ John Tradescant Jun.ᵗ
in his Garden.

4 *Welwitschia mirabilis:* a vegetable monster from Africa, ugly, unique and long-living. It so fascinated Sir Joseph Hooker that he examined its peculiarities through a microscope for a total of three days and three nights

2 *above left* The younger John Tradescant, like his father, King's Gardener. He was a searcher for 'curious greens' in the colony of Virginia and carried from the New World to the Old such familiar plants as Cornflowers, Lupins, Michaelmas Daisies and Bergamot

3 *left* Captain William Dampier: hydrographer, navigator, author, pirate chief and botanical collector

5 *Nepenthes rafflesiana:* one of the carnivorous Pitcher Plants native to the Eastern Tropics. To insects the pitchers look like elaborate banqueting halls; in fact, they are charnel houses. Such flesh-creeping hothouse plants delighted the Victorians

6 *right* Philipp Hans von Siebold, the Bavarian eye-doctor and botanist who introduced so many plants to the West, was twice removed from Japan. On the first occasion he was banished for having bribed the Imperial Astronomer to part with a secret map; on the second he was tricked into leaving the country by the Dutch East India Company

文政九戌年三月下旬未朝阿蘭陀人醫師シイボルトノ圖廿四歳
仙醫術ニ長モ候本年ニ至テ、ビルヘル、石案ヲ好ミ笠根ニ干磁石ヲ摺得シ

エリマキ　白木綿
ジユバンハンエリ　白木綿
下着　モ□レ□
上着　銀糸織人横嶋綢牡丹ガケ
　　　花色ラシヤ胸牡丹ガケ
下肢引　白メリヤス
上肢引　黒ジユス

山石崎常中正　寫生

7 *top* An expedition artist's impression of Commodore Perry's black warships lying off Japan in 1853. He carried a letter to the Shogun from the United States President which forcefully suggested a commercial treaty. This resulted in the opening of trading ports and other concessions being reluctantly granted to the Westerners

8 *bottom* Another artist's impression of an incident in the Chinese wars. The 1860 peace treaty which followed gave Western plant hunters the right to penetrate to the interior of China

9 A characteristic view of Szechuan, the 'Land of the Four Rivers' and
E. H. Wilson's special collecting territory

10 Some of the collectors organized by George Forrest. In twenty-eight years
he shipped home no fewer than 31,000 herbarium sheets and an equal number of
seed parcels

arriving safe and sound, the rate of plant mortality was increased by uninterested and idle sailors. Shifting the boxes to scrub the decks beneath was a deal of extra trouble and unlikely to be done unless the bo's'n happened to be by. It did not need much salt to ruin the plants. Seawater sloshed carelessly against the boxes day by day was more than enough. In the torrid zone it was vital to air the plants by lifting the lid of each box for a specified length of time, but in the torrid zone it was natural for seamen to do as little as possible. The least exertion increased their discomfort. *Ergo*, the lids were often untouched.

Rats belong to ships as eggs belong to bacon. Notoriously they will eat anything, and certainly they would not jib at the fresh wood of plant chests nor the salading which grew inside them. On the occasions, too, when insects were shanghaied from their native land on the leaves or stems of plants, their meals during the voyage were to hand.

With all these avoidable hazards was the shock to plant systems caused by the continuing variations in climate. A specimen collected in the high forest lands of Eastern Asia, and intended for Europe would have an immensely long passage and meet great heat, both humid and dry, and, worst of all, surprisingly rapid changes of temperature and humidity. Such changes which vexed the human spirit, were frequently the kiss of death to plants. Finally, in the storms which lashed shipping round the Cape and the Horn and on the last run through Biscay, the first thing to be jettisoned was always the consignment of plants.

All in all the safe transit and safe arrival of exotic plants destined for Europe was in doubt from the moment they were uprooted. John Claudius Loudon's calculation – and he was thorough with his figures – which gave the number of successful introductions, is both interesting and impressive. He noted that, at the beginning of the nineteenth century, 13,140 plants were being cultivated in the British Isles, of which only 1,400 were natives. The remainder, at some time or other and in various ways, had successfully made the journey from overseas.

The experiments carried out in order to lessen the mortality rate of plants in transit were novel and diverse. John Evelyn's advice was that plants stood a greater chance of surviving long voyages and the buffeting of foul weather at sea if they were stowed in barrels. Packing individual plants in paper wrappings and suspending them in nets from the main cabin roof was another proposal and it had a moderate success on brief voyages. On longer ones the plants were threatened by lack of light and moisture and the airless fug so much relished by seamen when they go below. Peter Collinson could only guess at Bartram's difficulties in keeping plants alive on his long expeditions into the interior; but he made some practical experiments in London and wrote to suggest the American put his plants in bladders, sprinkle a handful of moist loam over the roots, and tie the bladder's neck round the plant stem. The advice has a modern ring. How much easier Bartram's work would have been with a supply of polythene bags of varying size. Another of Bartram's patrons, the rich Dr Fothergill of Upton, made proposals for the effective shipment of propagating stocks; that they should be planted in open chests and protected from salt spray by canvas over a cane framework. In good weather the canvas could be folded back to let in light and fresh air. Alternatively the stocks could be planted in chests with removable lids and if the lids were glazed with talc or thick glass so much the better. Within the limitations of the day this was sound scientific advice, and Fothergill proved his wisdom by adding that the most important thing was to bribe captain, ship's officers and crew.

John Livingstone, the Chief Surgeon in Canton who was Reeves's friend and Kerr's detractor, wrote about plant transport in a letter to the Horticultural Society in 1819. Characteristically – for his temper was sharp and scolding second nature to him – he complained of this and that; but the letter was sufficiently informative and useful to be printed in the *Transactions* for distribution to fellows. His observations on the actual cost of shipping plants from China are astonishing:

> From my own knowledge and observation ... I am of the opinion that one thousand plants have been lost for one which survived the voyage to England. Plants purchased at Canton, including their chests and other necessary charges, cost six shillings and eight pence sterling each, on a fair average; consequently every plant now in England, must have been introduced at the enormous expense of upwards of £300.

At such a price it is something of a wonder that contemporary Al Capones and Giulianos failed to have a hand in so profitable and manageable a market.

It was natural that Livingstone should find the whole thing deplorable. He made recommendations based on observations he and Reeves had made in Canton. Because plantpot gardening was customary in the Orient the nurseries carried large stocks. But many of the plants were forced, which gave them sufficient vigour to last their time in a garden before being replaced, but quite insufficient to stand a long journey to Europe. Other, more spunky stocks ought to be obtained or specially grown. The growing medium was important, too. Nursery stocks in Canton were generally planted in local clay, which was excellent for their needs in the violent rains and long droughts common to that part of China, but inappropriate for a sea voyage. A more fibrous, loamy medium would have to be used, and plants settled in their pots a good two months before being put on board. A third important recommendation which broke with long tradition was that plants in transit should not be taken ashore at the island of St Helena. This custom may have been originated by a sentimental and unscientific sea-captain who, aware of the benefit a run ashore was to his crew, considered it would be equally beneficial to the plants he had on board. In fact the mild, moist climate woke up the partially dormant plants and started them into growth, a dangerous procedure which leached the reserves of strength they needed later in northern latitudes.*

There is no doubt they were excellent recommendations. But, insisted Livingstone, blowing clear, single notes on his own trumpet, this 'more certain method of gratifying the English horti-culturalist and botanist' would also fail, and all his own research, ingenuity and competence in the understanding of plants be entirely wasted, unless something was done about the appalling inefficiency at the Port of London. Jacks-in-office who held on to cargo manifests and slapdash stevedores could, between them, ruin a valuable consignment – and they often did.

It followed that trouble should be taken by the Horticultural Society to speed up the unloading and clearing of plant consignments at the docks. It fell to the Society as well to provide a full-time gardener in the East to undertake the selection of good plant stocks and their hardening and acclimatizing in

* On one of his two journeys to England Reeves demonstrated the worth of all three recommendations by successfully transporting 90 out of 100 plants from Canton to the Thames.

preparation for the voyage west. Scrupulous in detail, Livingstone counselled:

He should reside at Macao, having a suitable establishment, a house, garden, and native assistants.

And, as a spur to keep the man on his toes, he should be paid by piece-work:

... a certain handsome sum for every new plant with which he enriched the horticultural and botanical stores of England.

These last were, perhaps, counsels of perfection. Certainly in that category lies his recommendation that collections on ship-board should be put in the sole charge of a gardener. Yet only by careful application of the principles he advocated could the rate of plant mortality in transit be reduced.

Fourteen years after Livingstone's letter was received in London, Nathaniel Wallich, another botanist, who was a Foreign member of the Horticultural Society, sent in a paper on the same subject. His chief experience had been in transporting plants from India and, while generally underlining the recommendations from Canton, he particularized in practical matters. Plants, he advised, should always be mature – 'already advanced in age' – and grafts should have been firmly established for at least three years before consider-ing them for shipment. They should then be well established in square wooden boxes and placed in a glazed chest; eight boxes fitting snugly to each chest. This allowed for easy removal and replacement in case of deaths in transit. The little boxes should have three drainage holes bored in their bottoms and stand on a layer of broken glass and pebbles. The containing chest, on the other hand, should stand on small feet. This would sweeten deck-hands who could swab underneath them, and because the chests were painted, iron-bound and really the product of cabinet-makers rather than joiners, their elegant appearance would please captains who 'are very unwilling to allow their deck to be occupied by unsightly objects'. Air and water would be provided for by opening the cases at dusk and shutting them at dawn and by allowing one pint of water to each plant a day, half a pint in the forenoon and half at night. Although glazed with Chinese oyster-shell, stout glass, or plating of thick talc, the rows of chests were to be pro-tected from flying spray with painted canvas. These elaborations had proved successful, as Dr Wallich observed:

Such has been the vigour with which plants thus treated grew, that it was frequently necessary to have recourse to the knife in order to check their luxuriance.

As a description of plant transport in 1831 it sounds over-sanguine; wishful-thinking rather than statistical fact. Only two years afterwards *The Gardener's Magazine* was congratulating the purser of an East India merchantman for successfully carrying 8 out of 29 Chinese Azaleas from Canton to London. Being congratulated for a less than 50 per cent success seems a more accurate picture.

Because at the present time of quicksilver communications more seeds are transported than living plants, and because a seed's obvious advantage over a living plant is the small amount of stowage space it needs, it seems surprising that more plants were shipped than seeds until 100 years ago. Some collectors of the quality of Bartram and Douglas collected more seeds than plant stocks, but the very first record of a large-size seed collection being made was on the mortifying Amherst embassy to the Emperor of Cathay. Dr Abel's assistants then boxed up 300 lots of seeds, though shipwreck prevented most of them being germinated under the expert supervision of Sir Joseph Banks at Kew. This was in 1815. In 1822 John Lindley, then Assistant Secretary to the Horticultural Society, read a paper on the transport of exotics in which he drew attention to the value of seed as a means of introducing new species. To the layman it seems an obvious thing to do. But apparently it was not. As late as 1849 Joseph Dalton Hooker collected in the Sikkim Himalaya and after three months dispatched to his father at Kew no less than eighty cartloads of plants.

The general reluctance to collect seed may have been because of the difficulties encountered in learning the techniques of growing perennial exotics from seed. Today there are botanical laboratories where important data can be worked out: a seed's thermal constant, for example, and its proneness to pythium, the exact nature of its ferments, and the collection's 'percentage purity'. In 1849, and for a long time after, the importance of such work was not recognized. In any case there seemed to be little opportunity to undertake it for a number of botanic gardens were so understaffed or inefficient that huge backlogs of other work arose. The Musée d'Histoire Naturelle and the Jardin Botanique in Paris were inundated with collections of dried plants: for example only four of the many French missionaries sent in more than 233,500 between them. Yet more modest collections were also neglected. Seventy years ago Emil Bretschneider, the painstaking and wonderfully accurate author of *European Botanical Discoveries in China*, had grounds for complaint. He collected 'ripe seeds of interesting plants unknown in Europe' in the mountain regions of the Peking prefecture. These he 'transmitted for cultivation to several of the great botanical and horticultural institutions in Europe and America'. The results were not altogether satisfactory. The Jardin des Plantes and the Arnold Arboretum won his approval for the care they took, but he was scathing in his view that there was small point in sending seeds to the Royal Botanic Garden at Edinburgh owing to the poor treatment they received. It was fighting talk at the time, and yet no one of consequence appears to have contradicted him. Exotics were undeniably difficult to raise from seed. Without data from botanical experts professional gardeners must have often met with disappointment, and they would have far preferred to buy a consignment of plant stocks.

It is evident, also, that the carriage of seeds was never easy or certain of success, and though the collecting and packing of stocks was physically demanding it was more psychologically satisfying than collecting handsful of seed. Moreover the peculiarities of many seeds made absolute reliance on them something of a gamble. There were seeds as fine as dust and seeds like cannon balls; subfusc seeds and seeds as bright as jasper stones; seeds of a regular form and seeds from fantasy, shaped as parasols, as fans, and as little girls in party frocks. And some plants were reluctant seeders; others as prolific as spawning eels; and the cryptogams produced no seeds at all. Then there was the mystery of viability. The short life of Rubber Plant seeds was demonstrated by the need to charter a steamer to race them from Para to Kew; yet it has been said (probably apocryphally) that corn sealed in the Egyptian Pyramids reacted to moisture, warmth and light, and germinated. Some seeds which have been cared for with all the fancy aids of modern science have died quickly inside their cases; yet a leathery Tonkin Bean used for years and years to keep snuff fresh has gaily produced plumule and radicle when moved from the snuff box.

Idiosyncrasies of such a kind doubled a collector's anxieties. They were tripled by the fact that he was seldom in a position to gather seeds at precisely the right moment; and quadrupled because he was too mobile to make elementary germination tests on absorbent paper, felt or flannel, let alone the desirable but complicated tests involving the use of diaphanoscopes, separators, incubators and chemical balances with minute weights. Finally the collector's anxieties were quintupled because he could not even be certain that, even if the seeds did make the journey and germinate into seedlings, they would be true to the parent flower. Rogues are produced less by wild species than by cultivars, but they nevertheless exist.

In such a dense fog of anxiety, doubt and difficulty, it is no wonder that plant hunters were prejudiced against relying on seed as a means of introducing exotics.

There were also problems attached to the successful carriage of seeds. They were quite as large as those of transporting plants. Vermin relished nutty seeds far more than leaves and stems and roots. Changes of climatic conditions were also harmful to seeds. In great heat they over-ripened, which lessened their chance of germinating and, in any case, slowed up the whole process. A humid atmosphere softened a seed's cuticle and started a change in its reserve material; precocity which led generally to decay. The recommendation that bags of seeds should be suspended from cabin roofs was all very well, but the cool, drying air favourable to their health was rare. Conditions below decks aboard the old timberships varied between icy draughts and airless fugs.

Dr Fothergill, who advised Collinson and Bartram on the subject, suggested that large seeds should be crated in beeswax, set in a chipbox like potted shrimps in butter, and the outside of the box smeared thickly with a solution of sublimate of mercury. Smaller seeds, he said, could be wrapped in waxed cotton and treated in the same way. Alternatively, small seeds might be wrapped in screws of waxed paper and packed into glass bottles. But, warned the doctor, it was not sufficient merely to cork the bottles. They ought to be insulated against extremes of heat or cold by being cased in bladders or skins of sewn leather, and packed in a keg with a measured mixture of sal ammoniac, saltpetre and salt. No less elaborate was his third recommendation: to place packets of seeds in jars or canisters between insulating layers of rice, millet, bran or maize and top the lot with an anti-insect preparation of camphor, sulphur and tobacco.

Dr Fothergill's advice was hardly bettered for a century. Then the ingenious Dr Livingstone experimented in Canton with drying seeds artificially using one of Professor Leslie's ice machines which worked on the action of sulphuric acid. He was also an advocate of temporarily mummifying short-lived seeds in sugar – especially those like the Barberry where the removal of the fleshy fruit to reveal the 'stone' appears to shorten the latter's viability. His sugared seeds were put aboard Indiamen at Canton to be woken like a regiment of sleeping beauties on their arrival in Europe. And it was Livingstone who suggested that seeds which were so short-lived that they would die on the briefest voyage should be sown in pans at Canton and nursed through their germination by a capable gardener so that they arrived in Europe as robust seedlings. Whether the experiment was made is not known. The Horticultural Society certainly accepted his recommendation and passed it on. But delicate seeds and germinated seedlings would have to survive all the dangers which even Dr Wallich's three-year-old healthy stock found it an effort to withstand, and it was unlikely that until 1834 it would have been a practical possibility.

The year 1834 was calamitous in that it marked the premature death of David Douglas, the greatest of all collectors in the history of plant hunting. But it was also an *annus mirabilis* for botany, because in that year an experiment was carried out which proved conclusively the enormous value of a discovery made by a London physician, Nathaniel Bagshaw Ward. His discovery was accidental, but it so revolutionized the transport of exotics that plant hunting up to 1834 might appropriately be called pre-Wardian, and the intensive collecting done afterwards, post-Wardian.

Nathaniel Ward of Wellclose Square

Not a great deal is known of Dr Ward save that he was an amateur naturalist who practised medicine by necessity rather than choice in the East End of London. By deduction from the book which he published about his discovery we may guess he was gentle, probably shy, and acute but inclined sometimes to be pompous. Being the kind of man who puts caterpillars into bottles he was, maybe, either a bachelor or a widower or beshrewed. Certainly he was a romantic with characteristic longings for *rus in urbe*, and he recommended the invention he made because of its power to 'induct us into the quietest regions of vegetable life'. It seems improbable that professionally he was a ball of fire; more likely a modest, retiring physician in general practice. Within reason we may see him as a scientific cenobite or a recluse.

All this is conjecture, and is probably quite wrong.

What is known is that Wellclose Square, that part of dockland where he lived, was a Sherlock Holmes sort of place; not exactly producing lepers, abominable Lascars and wicked Chinamen, but giving that impression all the same. And had Holmes and Watson been acquainted with their contemporary, Dr Nathaniel Ward, undoubtedly they would have admired his scientific method of observing and deducing.

What Ward saw in 1827 was by chance: just as by chance the Curies saw pitchblende through their microscope, and Fleming saw by chance the fungus which led him to penicillin.

As was his habit Dr Ward had placed a caterpillar to pupate in mould at the bottom of a glass jar. He stoppered the jar and then apparently forgot it. When it came to his mind again he found that a tiny fern and an equally tiny blade of grass were growing out of the mould. This would have been an exciting event in itself in a part of London where greenery was as rare as hair on an egg, but the circumstances of the plant growth were of especial interest to the doctor:

> I observed the moisture which during the heat of the day rose from the mould condensed on the surface of the glass and returned whence it came, thus keeping the earth always in the same degree of humidity.

The cycle had, of course, been noticed before by glasshouse gardeners. Some had discovered it accidentally through forgetting to ventilate their houses on a hot summer's day. Others had deliberately applied the principle to make the warm, moist habit relished by cucurbits and most tropical exotics. But it required Ward's deductive powers to reason why the grass seed and the fern spore should shoot and grow into healthy plants. He concluded it was because they were in 'a moist atmosphere free from soot or other extraneous particles', and had light, heat and moisture. He added that it was also because they had 'periods of rest and fresh air'.

It is not easy to understand exactly what he meant by 'periods of rest'. Any kind of forcing, either stunting or hastening into precocious growth, would obviously be undesirable. And, left alone, plants rest of their own accord; they do not need 'shutting up' like parrots under a green cloth. Nor is it clear why changes of air should be so desirable. Ward's scraps of fern and grass thrived and gave an appearance of being immortal simply because he left the stoppered jar alone. He failed to state the fate of the pupa, which presumably perished, but he noted that the plants lived on in the same atmosphere for four long years – until the lid rusted and rainwater seeped in carrying with it the chemical impurities of Wellclose Square. Fresh air, therefore, would need to be given very sparingly indeed; firstly because it would upset the regular cycle of respiration, temporarily halting it and putting it out of gear; and, secondly, because the freshest of air might contain invisible and insalubrious impurities. It was in fact a feature of his closed cases that they were almost entirely glazed and were kept shut, or only opened on the rarest occasions. The closed cases used on sea voyages in pre-Wardian days were inadequately glazed, and either they were permanently closed, in which case the plants suffered from loss of light, or they were opened as regularly as weather conditions allowed and salt and other damaging particles were admitted with the change of air.

Developing his idea that a clean, even and undisturbed climate was vital to the health of plants, Ward designed a large-scale adaption of his caterpillar jar; instructing the carpenter who made it up that the joinery must be as near perfect as possible and the wood hard and thoroughly seasoned to withstand the effects of condensation from within and 'extraneous particles' from without. This case, in which he grew plants experimentally, was the proto-type of the hundreds and thousands which were to be made in succeeding years, for an important side-fruit of his discovery was that, in an age of gas lighting and heating by coal fires, indoor gardens became a possibility.

The cases were called *terraria* and soon they were as indispens-able a part of the early Victorian parlour and drawing-room as the velvet-hung mantelshelf, waxed fruit under domes of glass, 'Markart bouquets' made of dried grasses and peacock feathers, and occasional tables laden with *bijouterie*. Generally they were rectangular, though collector's pieces were made in round and oval form, wedge-shaped like cheese-dishes, and in the shape of urns and vases. In bizarre specimens the glass was yellowish or faintly blue, and sometimes engraved with arabesques and the profiles of con-temporary heroes. Cabinet-makers produced ornate, bow-fronted cases with reflectors and prisms on turntables to catch sunbeams, and their frames and stands were massive and made of carved mahogany and teak. Other cases were less ornamental but still solid and built to last. The simplest had cast-iron frames man-produced by foundries from a limited selection of moulds. These were fitted with containers like old-fashioned electric accumulators and stood on turntables covered with red plush or on stands with legs which varied in style from florid Hindu to elegant Regency.

The contents of the cases varied less. Egregious decorations such as chunks of tufa, sea-shells, stalactites and blobs of coloured glass were fairly rare. The plantings hardly differed in any of the thousands of indoor gardens then kept by the English-speaking peoples, for it so happened that the production of Ward's *terrarium* coincided with a public passion for growing ferns in both England and the U.S.A. It was a fern age. Ferneries, outdoor and indoor, for stove and temperate species, exotics or natives, for abnormal or regular varieties were exceedingly commonplace. Plant tinkers, called Botany Bens in England, scoured the countryside and carried baskets of ferns from door to door. Breeders and selectors spent a prodigious amount of time and money in raising crested, tasselled, foliose and plumose forms of common ferns until over a thousand varieties had been recorded and described. There were, for example no less than sixty-five varieties, hybrids and cultivars of the common Lady Fern. Inevitably the ubiquitous fern had pride of place in Victorian *terraria*, and being indifferent to neglect, resistant to most diseases, and as fond of the gloom as troglodytes, it was the easiest of all plants to grow.

The widespread indoor culture of cosseted Spleenworts, Rustybacks and Maidenhairs was, though, only an incidental result of Dr Ward's clever piece of deduction. The phenomenon he had seen told him this might be a means of meeting the difficulties of plant transport so frequently bewailed in the transactions of learned societies. He had this in mind when he designed his first prototype. Successful experiments led him to make other closed cases and try more ambitious proofs. In the year 1834 he proved conclusively that plants stood the highest chance of surviving if they were transported in sealed, glazed cases modelled on his caterpillar jar.

Two of Ward's cases were planted with English ferns and grasses, sealed, and fixed in a suitable position on board a clipper outward bound for New South Wales. After a six-month voyage the case was taken ashore at the Antipodes and unsealed. The contents were in first-class condition. Half the experiment had passed satisfactorily. By the doctor's instructions the two cases were then stocked with native Australian plants, and not ordinary ones, but plants which had already proved to be bad travellers. They included a species of the small creeping tender fern Gleichenia which had never survived any attempt to ship it to England. Ward was determined that his cases should be tested with exceptional thoroughness. Stocked with plants notorious for their tenderness and their reluctance to leave their natural habitat, they were sent on the long storm-racked voyage round the Horn. Between Botany Bay and the Pool of London the plants were subjected to variations of temperature from 20° to 120°. They were rolled sideways and tossed backwards and forwards on the swell and the roll of two oceans. But, at the end of it all, when Dr Ward went down to the quayside and opened his cases he found the plants secure, fresh and green, and full of promise. His confidence in the cases was entirely vindicated.

Thereafter the plant hunter's transport troubles were largely over; though mistakes, of course, occurred. It was possible, they discovered, to 'drown' a plant in its own respiration. Given a blazing sun and an excess of moisture in the case it was even possible to boil a plant to death. Inexpert joinery would admit salt. Unless an unbreakable transparent substance was used for glazing there were obvious hazards on a busy deck. Modifications were necessary and they were made, but the use of Wardian cases – as they came to be called – on overland journeys and long sea voyages meant that at last tender plants stood a good 90 per cent chance of survival.

It is rather natural to think of these cases as of a standard size under patent. In fact they varied between portable cases roughly the size and shape of a shoebox to the huge cases used by Joseph Hooker when he sent a consignment of trees from Tierra del Fuego to his father at Kew. These weighed upwards of three hundredweight each.

It is doubtful if Nathaniel Ward benefited financially from his invention, though a side perquisite was the sale of his book *On the Growth of Plants in Closely Glazed Cases*, a blueprint to the design and management of *terraria* and Wardian cases. But he was a man of science who had enough to meet his needs and he valued more than anything the congratulations of sages and the eminent. The gruff Duke of Northumberland, who had an Aladdin's Cave of exotics at Syon House, sent word to Wellclose Square that stove gardeners owed the doctor a great deal. His peer and horticultural rival, Charles, Sixth Duke of Devonshire, informed Ward that his case had been indispensable in carrying a specimen of *Amherstia nobilis* all the way from India to the special glasshouse made for it at Chatsworth. Sir William Hooker made a point of telling the doctor what a vast difference his cases had made in the transport of plants to Kew. Distinguished nurserymen like Conrad and George Loddiges of Hackney Wick found that using Wardian cases so greatly reduced their overheads that they were moved to thank the inventor. Other nurserymen, such as the firm of Veitch in Exeter, now considered Dr Ward's cases had made it worthwhile to train and send out plant hunters on their own account. Ward was especially delighted to hear that by means of his invention the

Cinchona Tree had been successfully carried for the first time from its native South America to Java, India and Ceylon, for as the only source of quinine the Cinchona was vitally important to the health of Asia. The East India Company was also very soon in the doctor's debt. Experts had advised that conditions in the Himalayas were wonderfully favourable to the growing of fine Tea, but the short-lived viability of Tea seeds had made it impossible to introduce the Chinese plant. Now, after Dr Livingstone's suggestion and under the expert management of Robert Fortune, seeds were sown directly into Wardian cases which germinated during the voyage, and over 100,000 vigorous seedlings were planted out in High India. The worldwide game of general post in many economic plants was only made possible by using Wardian cases. Seeds sent to Kew by consular officials were raised as plants and hardened before being consigned abroad again: Cork Oaks, for example, to South Australia, better Tobacco to Natal, and plants of Coffee, Cinnamon, Ginger and Indigo to Queensland. The movement of plants from one region to another in the rapidly expanding United States was also helped by Dr Ward's discovery. Undoubtedly it made dramatic changes in the prospects of plant hunters and their patrons.

After 1834, in the post-Wardian era, limitations seemed to disappear. Everything was now possible, or appeared to be. At last the whole world had become the plant collector's oyster.

PART THREE

A Scramble For Green Treasure

i

From Mexico to Patagonia

A Catalogue of Dangers and Privations: the Mounting of Expeditions: Joseph Hooker's 'Antarctic Herborizations': Veitch's Men – William Lobb and Pearce: Unlucky Collectors in the Cordilleras and Selvas: Orchid Stories – the Holy Grail of Orchid Hunters: Benedict Roezl from Old Bohemia

ii

In India and Africa

Opulent Attractions: the Kindly Dr Wallich: John Gibson's Career: Sir Joseph Hooker takes the Mantle of Sir Joseph Banks: His Adventure in the Himalaya and Assam: the Seductions of Dark Africa: the Welwitschia Affair: 'the Savage Life Again'

iii

The Eastern Tropics

More Veins of Orchid Gold: Thomas Lobb and Charles Curtis: Extensive Glass-gardening in the West: Beauties, Curiosities and Flesh-Creepers: the Sensational Pitcher Plants

iv

The Land of the Rising Sun

Shoguns and Mikados: Western Gunboat Policies: the Return of Philipp von Siebold: Less Privileged Collectors: the End of the Radical Revolution: Charles Maries, Sargent, Faurie and Japanese Plant Hunters

Post-Wardian China

The Opium War: Fresh Concessions to the West: Bentham's
Flora HongKongensis: Fortune's Four Trips; his Introductions;
his Prowess as a Tea Smuggler; his Indifference to the Taiping
Rebellion: Two Influential Botanists – Hance and Maximowicz:
the 1860 Treaty Opens a Way to the Promised Land

From Mexico to Patagonia

Masters of modern prose have averred that if they are used sparingly platitudes, like split infinitives, can have a special force. It would, then, be reasonable to say that plant hunting almost anywhere in Central and South America has never ever been a picnic.

Since 1492 the Panama Isthmus, the West Indies and the Caribbean Islands have suffered constant excitements, varying from the bloody raids of buccaneers and the suppression of slave revolts to more recent and melodramatic upheavals in domestic politics, and even the stoning of the umpire at a game of cricket. The region has seldom been in a sufficiently settled state for the comfortable pursuit of botanizing. As for South America, its history, its peoples, and certainly its botany, have never been without mystery if not shocks.

The practical difficulty of getting plants down country to the coast has been touched on. An additional anxiety was that in a continent where revolutions were endemic and, in some parts, *coups d'état* unbelievably common, no plant hunter when he returned from the interior could ever be sure of the political *status quo*. This made for deep uneasiness. And the actual physical and psychological dangers which confronted collectors out in the field almost defied description. Some, though, in their letters or journals made the attempt.

It will be remembered that Philibert de Commerson enjoyed his jaguar hunt near Buenos Aires. Not so the nineteenth-century collector who complained that in some parts of the country there was a 'jaguar every yard'. Another inferred that jaguars, however lithe and large and fierce, were far preferable to the vexatious pests which plagued without mercy – clouds of cabouri flies which 'settle on any exposed flesh like house-flies upon jam', *bêtes rouges*, ticks and diggas which bored and burrowed under the skin, stinging marabuntas, ants, the vampire bats which settled on the unwary and sucked and sucked until they fell heavy and befuddled with blood to slither into the undergrowth; worst of all, the leeches – black parasitic worms which could penetrate cloth and knew their way into the laceholes of breeches, the eyeholes of leggings, and the laceholes of boots as instinctively as inebriates into their favourite bar. These unpleasant manifestations were also common in the Far East and other tropical jungles, but collectors reported that in South and Central America they seemed larger, more abundant and far more malignant. A French Orchid hunter wrote in a jeremiad of his misfortunes that none of the tree-climbers he had engaged would climb up into the dense clouds of stinging insects, nor even collect in the scrub, where 'it was impossible to put down a shilling-piece between the red spiders and other insects'. He did not blame them, he said, for some of the ants were two inches in length; but it meant the end of his Orchid collecting in that part of Brazil.

The vegetation was formidable, and particularly in the tropical primaeval forests. Knives, axes and billhooks had to be used to carve trails through the undergrowth. The dense crochet-work of foliage and hanging creepers reduced light and increased humidity. The tree canopy distorted noises, too: increasing their intensity at night when the jungle sounds were as regular and as deafening as the noises of a busy city, the chatterings and screeches, the warblings and cleaks broken occasionally by the terrible screams of a creature in agony; and muffling them by day when the long periods of deep silence were equally unnerving.

The general effect of trekking in the rain forests was loathsome: of hacking a way through slippery, often tacky leaves and tendrils; treading warily on the slimy, uneven ground; watching as carefully as the gloom permitted for insects, ferocious animals, and a hideous variety of snakes.

All these privations had to be endured in loneliness, for it was characteristic of the vast jungles that unless the expedition was equipped and manned on a princely scale plant hunters were dwarfed by their environment into feeling desolately lonely; and the atmosphere was so enervating that an extensive Orchid hunt deep in the forests was not unlike living perpetually in a menacing and murky Turkish bath.

Fevers were feared more than anything else, for the plant hunter was immediately at the mercy of his porters. The unsuperstitious minority of the Indians might remain loyal and nurse him back to health; the larger majority would be more inclined to rob and desert him, or barter him piecemeal to headhunters and cannibals for plugs of tobacco.

This catalogue of dangers and privations was thoroughly understood by every potential plant hunter in South and Central America yet such was the quality of the exotic flora that they still continued to arrive.

Not many went south from the northern continent. The United States of America was growing larger. Year after year, like a shaken kaleidoscope, it wore a different appearance as territories were occupied, conquered, ceded, bought, assigned by treaty or added by revolt. Inescapably the *Zeitgeist* was inward-looking. Botanists and plantsmen had their hands full in dealing with all the wild species indigenous to the new states. The adventurers had taken to exploring, fighting Indians, and the widening of frontiers by the force of arms. The bored and the scamps sailed from New Bedford and Nantucket or Mystic in Connecticut to join in the bloody extermination of seals and whales in the South Seas. At that time only a few devil-may-cares were prepared to consider plant hunting abroad.

From old Europe, on the other hand, a plethora of private enthusiasts and well-equipped professionals took passage over the South Atlantic bound for Buenos Aires, Rio, Pernambuco, Para, Kingston and Belize. They were attracted in such numbers by the reports of explorers on the amazing flora to be found in the grasslands and selvas, the cordilleras and the Gran Chaco.

Victor von. Hagen in his lengthy book on botanists in South America* gave a skeleton list of the main scientific expeditions undertaken by many nationalities between the Académie des Sciences expedition of 1735 and Richard Spruce's work for Kew in the drenched selvas up to 1866. It included many honoured names: Charles Darwin, Löfling who was a pupil of Linnaeus, Mutis, Humboldt, Wied-Neuwied, Eschwege, Orbigny and Tschudi. By no means were they simply plant hunters. So large was every

* Victor von Hagen, *South America Called Them* (Robert Hale, 1949).

subject, so wide every horizon that scientific explorers in South America would naturally collect and note everything they could. Nevertheless the quality of their plant collections and their estimate of how much remained to be discovered promised so much to botanists and professional plantsmen in the post-Wardian era that expeditions were quickly mounted and sent out.

Some were very lavish, the general tone being that of Mahomet going to the mountain, and the more opulent private enthusiasts were great sticklers for what was and was not done, and would dine in modified evening fig even in the torrid zone. But grandeur suited the age and was of great psychological importance. It was strange but true that many Victorian plant hunters in the middle of a desert or the depths of a jungle were better equipped with durable comforts than the tenants of most public housing schemes at the present day. The brandy tantalus, the solid oak travelling-desk and medicine-chest, the brass-bound ditty boxes, the heavy, hand-made scientific instruments, and the linen, plate and cutlery compensated for the terrible adventures which lay behind and the terrible possibilities which, maybe, lay ahead. There was no doubt that all this paraphernalia, though heavy, and his strict adherence to the code of the society to which he belonged, kept up the collector's morale and gave him, apparently, fresh supplies of stamina.

There were, of course, a few nonconformists like Emmanuel Thomas who, far from adhering to any code of behaviour, did precisely what he wanted and never washed or shaved if he could avoid it. All the same his skill and his vigour excited Sir William Hooker's surprised admiration:*

He is never fatigued tho' he carries a great tin Box upon his back as large, and very nearly the shape of a London milk-pail, and in his hand an instrument for collecting plants which resembles a North-American tomahawk.

The collectors sent out by nursery firms were well provided for. Their equipment, if not as luxurious as that of the private plant hunter, was equally sound. It had to be. Central and South America quickly taught botanical explorers that skinflinting did not pay,

* Sir William himself was strictly correct and observed the proprieties wherever he was. When botanizing he wore almost the same clothing as he wore at the university or at Kew; and, top-hatted, high-collared, and equipped with the fobs and seals and watchchain of his class, as well as a vasculum for plants, he covered gigantic distances on foot each day.

that, in proportion, small-scale affairs stood far less chance of succeeding than the massive expeditions favoured by rich amateurs.
Nurserymen felt bound to follow suit and be generous in the salary,
kit and expenses allowed to the collectors they commissioned. But
their policy was less altruistic than businesslike. It was a promising
period for plantsmen. The new nabobs were expressing themselves
and trying to impress one another with the splendour of their
gardens and their glasshouses. The demand for exotic plants was
heavy and constant and Dr Ward had provided a means for shipping
plants. It was a sound investment to train and send out collectors
though, as they had writs to spend very considerable sums if it
meant the difference between success and failure, they had to be
men of integrity and judgement as well as brave, knowledgeable
and practical, resourceful, stubborn and single-minded. Today
such prodigies are state-trained astronauts and exceedingly rare.
In the last century they appear to have been fairly common.

The first and most notable botanist to collect in South America
in the post-Wardian era was Sir William Hooker's son, Joseph. At
the age of only twenty-two he was appointed Assistant Surgeon and
Botanist aboard H.M.S. *Erebus* on Ross's Antarctic exploration
which found the South Magnetic Pole. Though still a sprig,
through his father he was already well acquainted with many of the
leading botanists of his day in America and England, and he became a life-long friend of Charles Darwin, whose *Journal of the
Voyage of the Beagle*, published in 1839, had excited him so much
that he always kept a copy under his pillow.

Ross's expedition with the two ships *Erebus* and *Terror* was
thoroughly equipped with naval dockyard efficiency, though
Hooker recorded that the official botanist appeared to have been
overlooked. His own supplies were meagre:

> Except for some drying paper for plants, I had not a single
> instrument or book supplied to me as a naturalist . . . not a single
> glass bottle was supplied for collecting purposes, empty pickle
> jars were all we had, and rum as a preservative from the ship's
> stores.

The commander was sympathetic and at his own expense provided Hooker with a cabinet. Furthermore he never failed to

encourage him all through the four-year trip. But Hooker's greatest advantage was his own enthusiasm and self-confidence – enough and more for twenty botanists – and he collected with great gusto and no thought of personal discomfort whenever he had the opportunity. It was on this apprentice run that he laid the found-ations of his pioneer work as a geo-botanist, for he collected in so many different places that he quickly learnt to realize the import-ance of recording the precise location and habitat of every plant he gathered.

After Ross had found the South Magnetic Pole in what was to be later called George the Fifth land, the *Erebus* and *Terror* made sail to winter north of the Antarctic Circle, and so Hooker was given an opportunity he had longed for, that of collecting in that region bounded by Graham Land and the Falklands, and the southern-most buttonhook of Argentina. Getting there, though, was a dazing experience.

They met gales of such force that eight-inch and ten-inch hawsers snapped like strings of raffia, and squalls which ribboned their stormsails and laid both ships over on their beams. They righted again and again to be swept on through ice floes, running the gauntlet of icebergs which bashed together and snapped apart, or ground together backwards and forwards, upwards and down-wards like chewing teeth. The barometer stood at 28·40. Twenty-three degrees of frost solidified wave crests and spray as they sheeted over the decks. In one night of tempest the two ships lost contact with each other. On another night the *Terror* was ablaze for two hours from a fire below. On yet another night a freak blast of wind caused them to collide on the weatherside of a cliff-high iceberg. Their shattered topmasts and icy rigging knitted together, and when the *Terror* was chopped free the *Erebus* was found to be temporarily disabled, her gig and quarterboats washed away, and her decks a pig's breakfast of broken spars and rigging. It was only by Ross's superb seamanship that the *Erebus* was cleared from the iceberg so that they could make storm repairs in the belly of that frozen hell.

Limp, battered, bedraggled and in need of extensive carpenter's and sailmaker's work, the ships made Berkeley Sound in the Falk-lands on 6 April 1842.

Young Hooker, having been exposed to a longer and larger succession of perils than almost any plant hunter in history, continued to be bouncy. He was avid to get ashore and begin plant hunting. Ross's instructions forbade it. The purser's party was to go first ashore and collect mailbags. Hooker begged the purser to bring him a handful of plants, and he waited impatiently dreaming of hardy exotics on the bleak snow-capped shores. The purser returned hours later, without any mail but with one plant for the eager botanist. It was a tattered Shepherd's Purse.

Undiscouraged by this beginning Hooker scoured the Falklands on an extensive plant hunt and though the flora was small there were some interesting grasses to collect, one of which, the Tussac, which resembled the spray thrown up by a bomb dropped at sea, was to be grown at Kew and eventually take well and prove very useful in the Shetlands. To his father's pleasure (for Sir William was a great muscologist) Hooker found a splendid and rich variety of mosses on an island by Tierra del Fuego where Banks had been with Cook. On Fuego itself he was as fascinated by the flora as Commerson had been, though on this occasion there were no offers of a pretty Fuegian girl for either of the ships' cats. His enthusiasm for geo-botany already aflame, he decided that Fuego was 'the great botanical centre of the Antarctic Ocean', and his discovery that Britain and Fuego had many lichens and flowering plants in common interested him at once in the possibilities of plant migration.

The botanical drawings he made in this chilly part of the South American continent and the written-up notes of his plant collecting there were later published in *The Botany of the Antarctic Voyage*. He returned to England even fitter than when he left it, but the expedition had been demanding. Collecting frozen lichens, for example, had sometimes involved prising tufts out with hammer and chisel or, if he had forgotten to take the tools with him, sitting on a lichen until it was thawed by the warmth of his fundament. 'Antarctic herborizations', as he called them, made for discomfort but when a friend suggested that he compensate himself by plant hunting in the opposite sort of climate he confessed that for the time being he would rather stay at home with his father at Kew, having 'no notion of jumping from cold to hot and cracking like a glass tumbler'. Moreover, being by no means ignorant of discomfort caused by heat, he asked: 'Have not you botanists killed collectors a-plenty in the Tropics?'

The plant hunters up by the equator at that time would have

echoed the same view. South America claimed few, if any martyrs to botany through cold. Her heats slew many.

William Lobb was one of the most distinguished professionals to collect in South America. He and his younger brother both hunted for the firm of Exeter nurserymen called Veitch. Thomas hunted in Java and has a place in a later chapter, while William, a gardener rather than a botanist, was sent out to South America. He had all the attributes of an empire-builder, and, in the pursuit of plants, risked his personal safety to the point of lunacy.

Arriving at Rio de Janeiro in 1840, he set about collecting in the hinterland. It was difficult country and largely unexplored. He roamed through parklands and the primaeval rain forests along the edge of the Brazilian highlands. Very soon he had sufficient healthy plants to justify all Veitch's faith in selecting him, and in the next four years he added to his discoveries on the Argentine pampas, the Andes, and in the temperate forests of Chile. It was in the last country that he found a plant which was to establish his reputation as an out-of-the-ordinary collector. In a province inhabited by the Arancanos Indians he recognized a tree which had been first introduced to England by Menzies, the Scots surgeon on Captain Vancouver's voyage of survey and exploration. Menzies had called it the Chile Pine, but his introduction had been limited to five saplings, and the specimens were considered more of a scientific curiosity than anything else. Lobb, as a gardener, judged that the tree could become immensely popular in Europe and that it was worth sending quantities of pine kernels back to England. He also changed its name, or rather named the genus *Araucaria* after the Indians whose hospitality he had enjoyed and who used the kernels as a fruit or ground into a meal. His instinct had been right. With botanists his Araucaria from Chile enjoyed fame as being the only known conifer from south of the equator to grow to timber size in a West European climate. With a wide gardening public it enjoyed fame and great popularity as the Monkey Puzzle tree.

After his exacting expedition Lobb deserved a long rest as well as his patron's gratitude, but Veitch was too delighted by the success of the enterprise to be patient. He was allowed a short holiday at his home in Cornwall and then dragooned into leading another expedition to South America. He was to search for 'the new and the novel' in Chile and especially in the Chinos Archipelago and in Northern Patagonia.

It was on this long and arduous trip that he found the Lapageria, an evergreen climber with clusters of rose-coloured trumpet flowers which became popular enough to cement his fame. It took its name from the French Empress Josephine who, in her mercurial career, had once been Josephine de la Pagerie, and it was an instant success in Victorian conservatories. In the next few years William found a plethora of worthwhile introductions: the Scarlet Holly, *Desfontainea spinosa*, a shrubby flowering Holly not often seen today; the apricot-coloured Barberry, *Berberis darwini*, with its great and valuable resistance to saltspray; and the Fire Bush, *Embothrium coccineum*, which won high praise from W. J. Bean of Kew who wrote: 'Perhaps no tree cultivated in the open air in the British Isles gives so striking and brilliant a display as this does.' Yet another marvellous discovery was the Flame Nasturtium, a perennial with a preference for moist peaty conditions and so hardy that it stood being introduced to the more parky parts of Scotland and the Blue Ridge section of North Carolina.

It was to the United States that William Lobb went after his tougher collecting work was done, to hunt plants in the Sierra Nevada and the benign climate of California before dying of paralysis in San Francisco.

Veitch, whose principal aim now was to find stove plants and suitable subjects for bedding out, continued to invest in collectors. One of them, Richard Pearce, the discoverer of the bright green Plum Fir, *Prumnopitys elegans*,* also introduced the first of the Tuberose Begonias, *Begonia boliviensis*. With other South American varieties, found at great heights above sea-level it was to be the progenitor of all Veitch's famous cultivars.

After a lifetime of excitement Pearce was lucky to die in his bed. Other stove-plant hunters in the Andes and selvas were less fortunate. One, David Bowman, was robbed of all his possessions, including his plant collections, just before he was due to leave for England. Dysentery or mortification carried him off in Bogota. Another collector, Zahn the fern expert, got himself drowned in Panama. J. Henry Chesterton, the valet turned Veitch collector, overdid things while hunting Orchids and succumbed in Colombia.

His was a not uncommon fate. There was a huge variety of

* Poeppig called it *Podocarpus andinus*, under which name it is still sometimes listed in the catalogues of nurserymen. Its other common names are Plum-fruited Yew and Chilean Yew.

Orchids in the rain forests of Brazil and Central America and they had such a commercial pull that hunters frequently risked and lost their lives in searching for new species. Stories abounded about their difficulties. They abounded as well about the Orchids. Being voluptuous in character, expensive to buy and to keep, and coming from such mysterious habitats, it was inevitable. Disappearing Orchids had a prominent place in these romantic stories; that is, a popular species which died out in captivity and somehow had to be replaced. A well-loved variety of Cattleya Orchid called *C. labiata vera* flourished for a time in European stoves; then the plants weakened, refused to respond to treatment, appeared to have no interest in surviving, and, like Victorian ladies in a decline, died out one after the other until only one single specimen was left. In that age of industrial expansion there was money and to spare to mount an expedition to gather fresh supplies of the Orchid in South America. But by then no one could recall where the originals had been found. There were no records; simply the one living specimen in the conservatory of an English manor house. Gardeners, botanists, nurserymen and expert orchidists consulted one another, but even while they were wringing their hands and wondering how to keep the survivor alive, an Act of God in the shape of a holocaust resolved their problem, and manor house, conservatory and Orchid went up in flames. Years passed. Suddenly, colossal excitement was engendered when an old record was found which stated that the original *C. labiata vera* had, in fact, reached England entirely by accident – as a shanghaied plant in a consignment of mosses and lichens sent from the Organ Mountains of Brazil. Expeditions large and small, sponsored and independent, scoured as much of the Organ Mountains as was possible. *Cattleya labiata vera* had become the Holy Grail of hunters. And they could not find it. Then, seventy years after its disappearance, it re-emerged in Europe; rather improbably as the corsage of a lady at an embassy ball in Paris. An Orchid enthusiast attached to the British legation saw the corsage. He looked once, twice, in fact as often as a man may with propriety at an embassy ball; and, guessing what the Orchid was, he doubted his senses. Somehow he got hold of the flower. An expert confirmed his guess. The Holy Grail had at last turned up; and the trail was followed back to the precise place where it had originated in Brazil and Orchid fanciers were able to obtain fresh stocks.

Some Orchid hunters had heroic adventures and endured privations which deserved to feature in popular ballads. Walter Davis, for example, crossed and recrossed the Cordilleras of the Andes a score of times, reaching the great height of 17,000 feet; and he travelled the length of the Amazon from its principal source to the North Atlantic delta. Gustave Wallis made the same mammoth journey in the search for Orchids but in the opposite direction, from mouth to source, and, unlike Davis who survived to return home, he was killed by a fever in Ecuador, shuddering his terrible way through to madness and death.

In contrast there were collectors who had a remarkable share of luck and more than their share of comfort. Albert Millican wrote up his experiences in a floridly-decorated volume called *The Travels and Adventures of an Orchid Hunter*. The initial letter on the first page is 'illuminated' by the then infant process of photography, and contains a portrait of the author. He appears to have been naturally cushioned against privations, far less like the lean and leathery plant of one's imagination than, for all the world, like Edward VII caught in a benevolent mood after eating lobster or winning the Derby. Millican's enterprises, though to some extent rewarding, made fewer demands on him than was usual in his profession.

Prince of Orchid hunters in Central and South America was a Czech named Benedict Roezl. The majority of his fellow collectors accepted him as such, and admitted his superior knowledge of the likely whereabouts of Orchids in the selvas. His 'bags' were richer than anyone's. He introduced different forms of Selenipediums, South American cousins of the Lady's Slipper; varieties of Cattleyas and Stanhopeas; and curiosities such as the Masdevallias he found which had long sepals trailing like kitetails, and the Miltonia which was called after him, *M. roezlii* – a tree-perching Orchid with the capacity of Paestum's famous Roses for flowering twice in a year. No one outclassed him as an Orchid hunter in the steaming rain forests which run between the Gulfs of Campeche and Guayaquil and stretch from Bolivia to the Atlantic coastline of Brazil, but Roezl won equally high admiration as a general collector, and in this capacity he made plant discoveries in places as far removed as California and Patagonia. They included some well-favoured Lilies from the wilderness as well as from the cultivated gardens of Spanish settlers; Agaves and other succulents from the Mexican plateau and the desert lands of Atacama and Patagonia; stove plants, such as the pygmy *Aphelandra aurantiaca* with its twisted leaves and surprising colour of a blushing Chinaman, and, from the West Indies, Flamingo Flowers, with spathe and radix not unlike that of a wild Parson in the Pulpit. There was *Anthurium roezlii*, for example, with its white pulpit and port-coloured parson; and others with parsons of ivory, vermilion and raw pink.

Such successes ought to have made Roezl a rich man. Joseph Hooker, when complaining of his own meagre government grant as a collector in Assam, wrote that 'an active collector . . . might easily clear from £2,000 to £3,000 in one season by the sale of Khasi Orchids'. He inferred that not many Orchids were necessary to raise such a figure. Roezl, although one of his consignments of Orchids from Central America actually weighed eight tons, never made a great deal of money. Probably this was because he seldom collected as the employee of a nursery. Working on his own account or on special commissions from private enthusiasts suited his temperament far better. He had been born near Prague, then the capital of Bohemia in the ancient Austro-Hungarian Empire, and, rather curiously, he had managed to inherit not one but both strands of the national character. To the stupefaction of his porters he would alter, chameleon-like, but for no apparent reason; at one place and time being thin-lipped, silent and wearing a purposeful, doomed, proud look, at another showing himself irrepressibly optimistic and sunny. He was a stimulating travelling companion, and a genius at tracking down plants, but too impulsive by far and too fiercely individualistic to collect under the direction of financial investors. Later, when he returned to Bohemia and set up as a nurseryman himself, he found trading in plants a fiddling affair unsuited to a man of spirit and his business failed. He was at his best in the roasted Mexican deserts, the swamps of Costa Rica, and the trackless forests of Colombia; modestly accepting the high regard of his fellow collectors, delighting in finding new quarries, and – depending on the mood of the moment – being either indifferent to hardship and danger or doing his best to laugh them away. He had quite sufficient of both.

Acclimatizing is an enervating and uncomfortable process. Roezl, shifting from climate to climate and altitude to altitude, had to have the stamina of a plant being carried to and fro between tropical stove, temperate house, conservatory and cold frame. If he did not penetrate that part of South America where the plant hunter had complained of 'a jaguar every yard', he had his share of close contact with them. On one occasion he was working in his tent, writing up notes on the day's collecting and pressing specimens, when he was made abruptly aware of a pungent cat stink. He turned to find a jaguar two yards away, and it was neither a kitten nor a benign old grandfather but middle-aged and very evidently disgruntled. It snarled, rubbed its head against the table, and clearly was considering what to do. But before it could make up its mind Roezl came to life, hurled an oil lamp at the beast and threw himself to the floor. The suddenness of the act, Roezl's speed, and perhaps his effrontery, made the jaguar turn tail and disappear with a whimper of fear.

If wild animals were a constant danger, no less so, in the half-civilized party, were the attentions of bandits. Roezl appeared to be bandit-prone. He was held up and robbed no less than seventeen times; though because he was one of the few collectors of the day to travel when he could both austerely and lightly, he never lost a great deal. One band of Mexican bandits were so exasperated to find he had barely anything except his horse, his clothes, and bundles and bundles of weeds that some of them were inclined to cut his throat. The knives were out when the chief held up his hand and bade them reconsider. Could this weed-collector be sane? They eyed one another. It was bad luck to cut the throats of lunatics. If a man was so crazed as to wander the countryside gathering flowers and grasses and bits of trees, he must surely be under the eye of God. This, anyway, was their happy conclusion, and, feeling the need of a little grace at that particular time, they piously crossed themselves, let Roezl go, and rode off to rob someone else.

It is clear that he had a Benjamin's share of good fortune. Another propitious sign of it was a heaven-sent gift of plants in 1871. He and a native boy were canoeing down to the Pacific coast after an unrewarding plant-hunting expedition up in the mountains. The river was swollen with recent rains and flowing swiftly. Alert to every danger as they rocketed downstream Roezl found they were overtaking a floating tree which must have been uprooted and torn from the bank far away up-river. Its branches and bark were festooned, as though for a floral carnival, with epiphytic Orchids, ferns and mosses of every shade of green, and (most astounding of all) a crimson Monochaetum with furry leaves. Roezl and the boy quickly moored themselves to the tree and collected plants from its branches as they travelled down-river, the rich haul completely compensating for the disappointments of their expedition.

Such luck, with a great deal of courage, the skills he learnt and acquired as a hunter, and his own special talent for nosing out rarities were at the back of his immense success. And they allowed him to survive each of the human, animal, insect, vegetable, geographical and climatic perils which he encountered so often and with such aplomb in South and Central America, and die, moreover, 'of natural causes' back at home near Prague. He merited the statue raised to his honour as the most eminent plant-hunting son of old Bohemia.

In India and Africa

From 1498 when Vasco da Gama cast anchor in the harbour of Calicat, India was like a candle-flame to moths in the minds of imaginative Europeans. She promised opulence. Faustus, tempted by the 'evill Angell' longed for spirits to 'flye to India for gold'. Moreover, she promised mystery, novelty and enormous opportunity to adventurers, explorers, traders, soldiers, empire carvers, administrators, religious proselytizers, civil engineers and, ultimately, to plant hunters. The land of Warren Hastings, Clive and Lord William Bentinck was also the chosen plant-hunting territory of several botanists and plantsmen.

The long-established rule of the East India Company and, after the transference of power to the Crown in 1858, the order imposed by British India, gave Anglo-Indians (and the term was not then abusive) a secure feeling of 'belonging' which settlers in Australia have yet to experience. There was no feeling of temporarily camping-out, and the more solid attributes of Western culture were soon in evidence. Amongst them were botanic gardens.

Nathaniel Wallich, who had made sage recommendations on the carriage of plants before Ward invented his case, was a director of the Calcutta Botanic Gardens. He was himself a collector of renown and his herbarium material eventually fetched up in Kew, but, like Reeves in Canton, his chief service to plant hunting was really the help he gave to fellow botanists and to gardeners sent out by the great. In particular he had the wit to guess the stresses under which the latter worked. To begin with they depended absolutely on succeeding. It was at the back of their minds that if they did not track down and transport what they had been ordered to find, their livelihood and their whole future might be affected. It was unlikely, but it could happen. They would also be homesick, and, in the class-haunted societies of the nineteenth century, they would be either too shy to ask for help from the upper-class English and French residents, or – made bold by necessity – they could so easily appear arrogant and overbearing in their demands. Plant hunting in India was difficult enough without adding an extra load of tensions such as these, and it was Wallich's special talent to help. It was he who met the gardeners personally at the dockside. It was he who smoothed out the difficulties, soothed the real or imaginary hurts, and made sure they were well looked after and respected all the time they were in British India.

Wallich's care of John Gibson, who arrived in 1835, that is, in the first year of the post-Wardian era, was illustrative of his general concern to be as useful as possible. The young man was sent to India by the Duke of Devonshire, or rather by his head gardener, Joseph Paxton; and his instructions were precise: he was to find *Amherstia nobilis* and as many Orchids as he could.

Gibson had been chosen because he was considered the Chatsworth 'intelligent gardener', but he was naturally predisposed to be over-anxious, and as his ship approached the Sundabans he was dismally convinced he would be an utter failure. No one had troubled to tell him very much about India, but he was familiar with what their local Derbyshire poet Erasmus Darwin had written about a Lotus floating on the Ganges:*

> Charm'd on the brink relenting tygers gaze,
> And pausing buffaloes forget to graze;
> Admiring elephants forsake their woods,
> Stretch their wide ears and wade into the floods. . . .

It was barely surprising that so extravagant a picture failed to correspond with reality, because Erasmus Darwin had never been to India; but the almost total lack of exotic beasts and, instead, the unromantic, dull network of turgid, brown rivers at the delta of the Holy Ganges, was something of a let-down to the country boy from Chatsworth. As he later confessed, when he stepped ashore he was disappointed and trembling with uneasiness. Depression had

* Derbyshire people were very proud of Darwin as author of the best-selling *Botanic Garden*. He was the owner of an eight-acre botanic garden, the writer of 'fantastical, extravagant often incomprehensible verses', and grandfather to the great selector.

drained him of all confidence. He could not bear the responsibility which faced him. His lavish equipment (no less than thirteen of the new Wardian cases), and all the vast Cavendish wealth which in theory lay behind him and which would have bolstered up most men, merely increased his gloom. The thought of Paxton's elaborate plans to construct a special stovehouse for the *Amherstia* unnerved him more and more. And he could not bear the heat. No one at Chatsworth had thought to inquire if he enjoyed hot weather. He did not. In fact he found it unendurable. A temperature of between 80° and 99° made him, in Wallich's words, 'dreadfully miserable!'

Wallich came to the rescue. He paid small attention to the gaucherie and the long silences of the young gardener. He charmed away his uneasiness; gave him the confidence he needed; promised him that though Bengal could be as airless and blazing as a smelting furnace, and Assam as sodden as a sponge, the plants he saw would give him such pleasure and interest that they would outweigh all inconveniences and discomforts. He was right about both. Carried by boat and wagon and then in a *palkee* by relays of men called his *dawk*, Gibson moved from stage to stage up-country through East Bengal and the valley of the Brahmaputra. Lying in the curtained *palkee* he found the climate unbearable and preferred to walk. However sizzling the tracks, however drenched the forests, his mind was distracted and he found himself spellbound by the vegetation. Where it was sparse it yet outrivalled in brilliance anything he had seen in the soft watercolour country around Chatsworth. Where it was lush, its density was extraordinary: thick choking masses of ferns and mosses, funguses and lichens, Bamboos soaring up to the height of a church tower 100 feet and more, trees rigged like shipping with shrouds and stays of rampant plants, epiphytes gleaming here and there on trunks and branches; nature so tumid that, in the rare moments of silence from the crying and chattering and calling, the shrieking and screaming, the yelping and roaring, the rate of plant growth, the building-up and burgeoning cells, seemed almost audible.

Gibson's self-doubts vanished. He found his *Amherstia*, and so strenuous and effective was his other collecting that when he sailed away from Calcutta all thirteen of his brand-new Wardian cases were replete with exotic plants. Good Nathaniel Wallich, the inspirer and encourager of his success, saw him safely aboard, helped to arrange the epiphytic Orchids which, still attached to their hosts, were suspended from Gibson's cabin roof, and then took the trouble to write to the Duke of Devonshire that his gardener's treasure of Indian plants was 'a truly princely one'. It was. The efficient Paxton, forewarned by Calcutta, had a ducal fly-boat waiting at the canal basin close to the docks where Gibson's collections were put ashore. From there they were raced by teams of horses to Cromford, the nearest canal basin to Chatsworth, and thence by sprung wagons. In record time the Duke of Devonshire's new Amherstia House had its first occupant. And besides this triumph which 'afforded the Duke great satisfaction', Chatsworth was the richer by more than a hundred species of Orchid and, as Gibson himself reported to Paxton, 'other fine plants which were not in England when I took my departure from India'.

The success of Gibson's first botanical exploration gave him the confidence to make many others, and he became renowned as one of the most experienced of all travellers in the Eastern Tropics. Because of this he was ultimately appointed superintendent of Battersea Park, which he laid out himself. By using foliage plants such as Fan and Sago Palms, Tree Ferns and Bananas and Aralias, which Sir William Hooker grew for him at Kew, he attempted for the enjoyment of all Londoners a reconstruction of the rich vegetation he had seen on his many travels. Very soon the Battersea Sub-tropical Garden was as famous in London as the exotic park laid out by Barillet-Deschamps in Paris and Olmstead's Central Park in New York. But to Gibson himself it was a pale imitation; merely the ghost of the dripping and matted jade forests enamelled with colours which he had first seen in Bengal and Assam under the aegis of the director of Calcutta's botanic garden.

Wallich was celebrated amongst his friends for what he called his 'little notes'. Any excuse to write one was immediately seized on. He would, for instance, present plants to local botanists and gardeners in Calcutta and then subject them to a bombardment of botanical *billets doux*, inquiring into their progress, rate of maturity, health and appetite as though the plants were children not vegetables. And he adored sending good news, writing 'little notes' of congratulation with boyish glee. A good example was one sent to Sir William Hooker when his son was elected a fellow of the Royal Society. By then Wallich had given up his work in Calcutta and was able to be present at the election. Barely had it finished before he was writing to Sir William:

I rejoice to tell you that your son was elected by a vast majority ... much greater than any among eight candidates that were successful. Your Joseph *beat them all*! Hurrah for your Name and for the cause of Botany!

Wallich's enthusiastic support was not only because he was a friend of the family but also because Joseph Hooker was about to go out to his beloved India to collect for Kew and the Geological Museum, and the young man already showed promise of being in many respects Elisha to the Elijah of Sir Joseph Banks. The promise was to be fulfilled. He was to be heaped with honours and elected President of the Royal Society as Banks had been. His influence was to be as extensive, his knowledge as large, his experience as a field-collector as wide and as rich. In the whole history of plant hunting there was never a pair quite like the two Sir Josephs.

But when he became a fellow of the Royal Society Joseph was only thirty and not yet knighted. Since his return from the 'Antarctic herborizations' he had been overfilling each day with a great number of projects: compiling his *Flora Antarctica*, drawing sets of plants for *The Botany of the Antarctic Voyage*, lecturing at Edinburgh University with a view to keeping the Chair of Botany warm for himself (though when it fell vacant he was not, in fact, appointed), refusing the Chair of Botany at Glasgow which was offered as a sop, and turning down as well the suggestion that he might like to be Director of the Sydney Botanic Garden, considering and then accepting a post as fossil botanist on the British Geological Survey, writing essays on Coal Plants, working on a collection of dried plants from the Galapagos Islands sent to him by Charles Darwin: all this, and falling in love with the daughter of the Cambridge Professor of Botany.

At this point in his life Joseph might have turned in one of many directions and settled to a distinguished career in this or that. But he could no more sit still at home than Banks before him, and he was no more a laboratory botanist than Banks had been. He had the itching feet of a field botanist, and so when on the Isle of Man he happened to meet the First Lord of the Admiralty, who rather improbably offered him work abroad, Joseph hardly paused to consider. The first proposal was that he should go out to Borneo and prepare a report on the British possessions there; and then, on his way home, collect in India. This Borneo project never came to anything. Collecting in India, though, became a firm offer – fossil plants for the Geological Museum and living plants for Kew. Though his writing would have to be neglected, and he would be separated from the Professor of Botany's daughter, Joseph had no hesitation at all in accepting the offer. It was what he really wanted. The results for horticulture and botany in the West were far-reaching.

Far better equipped than on his last plant-hunting expedition, armed with last-minute advice from Nathaniel Wallich and accompanied by Hugh Falconer, the new Director who was going out to take over the Calcutta Botanic Garden, Joseph travelled by ship to Alexandria, barge to Cairo, dromedary to Suez and thence by the *Moozuffer*, an intensely uncomfortable steam frigate. In no time at all he was working on the first part of his commission and collecting fossils. That done, he made for the Himalaya.

Gibson's pre-vision of India had been tinted by Erasmus Darwin's extravagant verses. Hooker's was tinted by an account of Samuel Turner's embassy to Tibet in 1783, and though he was an excellent and meticulous botanist and geologist he was also a dreamer. He was one of those rare and very contented human beings with a capacity for pleasure which included both the scientific and the romantic, and could enjoy the tiny detail shown through a microscope as well as the scenery which was sublime. The second part of his commission excited him. Sikkim was his objective and, if he could get there, the independent states of Nepal and Bhutan. From all he had been led to believe about the climate he stood a fine chance of discovering many unknown botanical species. This was scientifically satisfying. He was equally delighted by the romantic notion that he would be the first European for sixty-five years to travel in that unmapped region of the Himalaya; and, could he have known that half a century would pass before anyone else followed in his tracks, he would have been enraptured.

In the account which he wrote of his travels in *Himalayan Journals* he showed the bizarre chauvinism and arrogance of a contemporary Englishman abroad which once characterized the spread of red on maps of the world, and the memory of which continues to embarrass the over-sensitive, enrage the radicals, cause the breasts of those who have gunboat nostalgia to swell like frogs, and the puckish to chuckle. Hooker's generalizations were exquisite little miniatures of Victorian hauteur. The Lepchas he encountered were worthy of some praise, though he added the modification: 'I believe the Lepcha to be a veritable coward'. The Bhoteeas were 'queer and insolent'. As for the Khasia people, they were 'sulky, intractable fellows'. Still, he did his duty by them and politely accepted the honours paid him by villagers when they brought him gifts of fruit and vegetables, made sun-shelters where he could rest, and bored him with ceremonious beer-drinkings.

He travelled in some luxury, with a suite of anything between sixteen and sixty servants. This, though, did not necessarily mean that he was comfortable. The contrary was true when bearer replacements failed to turn up at an appointed stage and only after hours of extensive negotiations in teeming rain could he persuade six men to carry his litter while he himself walked the fourteen miles to the next stage. By no means, either, was this an isolated example of administrative inefficiency. On another occasion he had to spend a night in a hut without bed or food because some baggage coolies failed to turn up. Far worse, on his way to Bhomsong in Sikkim, where no European had ever been before, he was obliged to subsist for eighteen days on eight days' provisions, and was reduced at length to an exclusive diet of rice and chili vinegar. And, worst of all to a zealot like Hooker, because the coolies who carried stocks of drying papers sometimes lagged behind or, mysteriously, were missing when they were most needed, he could not always preserve the plants he found and had to throw away collected specimens. His chagrin when this happened can only be imagined. He was made testy, too, by local politics, which seriously threatened his work and even his liberty. In theory the Rajah of Sikkim was an autocrat. In fact, he was under two thumbs; that of his dewan, a Tibetan, who had a keen appetite for intrigue, and that of the governor-general, Lord Dalhousie, who had an equally keen appetite for annexation. The latter's political agent in Darjeeling met Hooker at Bhomsong and with some difficulty managed to present him to the Rajah. Thereafter, though he had not the slightest interest in the matter, Hooker was committed willy-nilly to be anti-dewan, and affairs reached such a stage that for no apparent reason he and the political agent were arrested and imprisoned. The agent, a mild-mannered doctor named Campbell, was put to the torture, and their fate hung in the balance for over a month. Then, by great good fortune, the dewan overreached himself in hatching plots and was disgraced, and Dalhousie was at once pleased to annexe Sikkim for her own protection.

Hooker's own report on this piece of history was modest and restrained. Having been released from gaol, he continued botanizing as though nothing had happened. He had his full share of that maddening English sang-froid which enabled him to face angry Sikkim sepoys, far less frightened by their hostility than outraged at their effrontery, and armed only with a foot-long barometer. And the four-year Antarctic Expedition had taught him fortitude. Without complaint he endured such excesses of Asian climate as storms of continuous lashing rain for sixty hours at a time, extremes of heat, cold, dryness and humidity, as well as the depredations of fleas, bugs, ticks, midges, and leeches which permanently scarred his legs. With the same modesty and restraint he mentioned, as though in passing, that in climbing to 19,300 feet he went higher than any record to that date. He was far more pleased to record his good fortune in seeing the Spectre of the Brocken – his own shadow cast on to mist with a rainbow nimbus about the head. He happened to be the first Englishman to feel the deep chill on the eternal snows of Kangchenjunga, to know the ceaseless and piercing winds which howled about its precipices, and hear the thunder of its glacial avalanches, but he made no more song and dance about it than in describing the large yellow-flowered Begonia used by the Sikkim Indians as an ingredient in sauce.

Hooker's *Himalayan Journals* show very clearly that his tastes were catholic and his scientific curiosity unusually keen. They do not emphasize though, that in India he was confirming by personal experience his important theories of distribution. The Himalaya were so immense that on their mountain slopes it was easy to see the variations in plant life at different heights above sea-level. His observations were confided, as ammunition for the *Origin of Species*, to his friend Darwin who, in his reply, correctly prophesied:

I know that I shall live to see you the first authority in Europe on that grand subject, that almost keystone of the laws of Creation, Geographical Distribution.

It was not only this which marked Hooker's Indian journeys with such interest. His collections in Sikkim and in 'leaning over the border of Tibet' between April in 1848 and November 1849 included twenty-eight entirely new Rhododendrons besides a vast quantity of genera and species. Vast is the only appropriate adjective. His consignment of plants collected in the three autumn months of 1848 amounted to eighty loads. In Assam, where he collected on his way home from Sikkim, he collected more than 2,000 flowering plants including Orchids, Balsams, Ferns and Mosses, as well as over 300 specimens of different timbers for his father's Museum of Economic Botany at Kew. Western gardens were considerably altered as a result of his introductions, especially in the moist peatlands which correspond most closely to the Khasi Hills of Assam. In Hooker's seven-month stay there he measured no less than 300 inches of rain, an inundation which seems even more impressive if it is written as 65 feet. Such Noah's-ark downpours are unknown in places like North Carolina and the Scottish Highlands where Hooker's Rhododendrons grow, but they are wet and peaty enough.

Three in particular of his many Rhododendrons have claims on the discriminating plantsman. *R. thomsonii* produces interest and colour from all parts like a good orchestra producing music. Its flowers are colossal and an unforgettable red, but it also gives pleasure with its foliage. The young leaves are frog-coloured when they replace the old, and afterwards they turn to a turquoise shade. And from the time the shrub is middle-aged its bark peels to reveal a medley of mellow colours from violet to bulrush brown. *R. nivale* is less spectacular: an arctic shrub appositely named the Rhododendron of the Snows. Hooker found it growing like Heather at 18,000 feet, and was the first of many to be enchanted by its little blossoms, the colour of Common Thyme flowers, and exquisitely fragrant. The third has a special place in the affections of colour connoisseurs. The foliage is large and a bright lettuce green, the flower bud a darker shade, about viridian, and the flower chalices are the palest yellow shot with a third distinctive shade of green. Under the complex filing systems of Rhododendron experts the shrub now belongs to the Series maddenii and the sub-series megacalyx, but it is better remembered by the name Hooker gave it to honour the seizer of Sikkim, *R. dalhousiae*.

Hooker was one of the few front-ranking botanists to collect in

Africa during the immediate post-Wardian era. Plant hunting there before had been in the capable hands of men like Commerson, Masson, Bowie and Burke. During the greater part of the nineteenth century, however, Africa was as dark a continent botanically as she was in most things and no extensive collections were made there in any way comparable to those made in Central and South America, Asia and the Far East. Africa attracted Hooker largely because she was romantically 'dark'. She contained great untrodden lands as well as neat gardens – some more ancient than many in Europe and in the most improbable locations. The Nefta Oasis, for example, either sizzled with heat or shivered with cold. At noon 135° in the shade was not uncommon and Norman Douglas, writing in Hooker's lifetime, noted that the Nefta Gardens could only be maintained through being 'assiduously tended and groomed by relays day and night'. And Africa's romantic attractiveness was undoubtedly sauced for Hooker when a Dr Welwitsch, botanizing in South Africa, came across the genus which bears his name, *Welwitschia*.

Depending on the point of view of the beholder the plant was either scientifically fascinating or a vegetable monster. Hooker naturally held the former view. It was ugly and unique and lived for centuries. Therefore its only known variety deserved the name they gave it, *W. mirabilis*, and it was not surprising that for a time it stole Hooker's heart away. Flowering cones and sketches of the plant were sent to him at Kew by the house-painter turned botanical artist, Thomas Baines. Not very long afterwards the real thing arrived from Dr Welwitsch. Its trunk was the shape of an inverted cone and from it sprang a pair of long, flat, leathery leaves and cymes of scarlet cones. By the time Hooker received this unlovely gift from Africa he was married, a father, appointed his father's assistant at Kew, and busy, as ever, with a dozen different projects, but the misshapen plant devoured all his interest and spare moments for months. Its peculiarities absorbed him. He examined the plant's parts with scrupulous care, an eye glued to the microscope in long vigils of five hours at a stretch, and for a total of three whole days and three whole nights. He lectured on it, talked about it, wrote about it and no doubt dreamt about it. Africa, its native land, would appeal to him enormously – especially as he was not by inclination a microscope botanist. He was as restless as any true traveller and confessed to the American botanist Asa Gray that he longed 'to taste the delights of savagery again . . . to get into the jungle and live in tents'.

Morocco, where he chose to go, possessed no jungle, but it was still one of the least explored parts of the globe. Hooker set off with two companions and a Kew gardener called Crump, and though the party was bedevilled with the need to bribe officials and there were annoying setbacks and delays, they combed the Atlas ranges and took home a fine collection of plants.

Hooker returned to a truly Banksian existence as the complete controller of Kew's fortunes, government adviser on botanical appointments, and President of the Royal Society. Later he enjoyed another stimulating expedition, this time in the Rockies with Asa Gray. And he was heaped with honours from universities, learned societies, governments and reigning sovereigns.

Enduring hardships aboard the *Erebus* and in Sikkim, the cold of the Sierra Blanca and the blistering heat of Morocco, evidently suited him. He lived on and on. In his great old age he was the benevolent despot of all botanists, and retirement from the responsibilities of administering Kew made little difference to his liking and his capacity for work. His last undertaking was self-imposed and one of the most arduous in all his career. With a microscope and herbarium sheets sent to him from Christiana, St Petersburg, Berlin, Paris and Kew, he set about the laborious task of classifying the *Impatiens* genus in the Balsam family. He managed it, naming 303 new species in the process, returned the herbarium sheets, put away his microscope, and laid down his pen. Four days later he died; it might be said of nothing in particular. He was ninety-four years of age.

The Eastern Tropics

The quantity of shipping controlled by Britain in the nineteenth century ensured that the Western plant hunter on his way out East could enjoy home comforts to a surprising degree. The British merchantmen painted in reds and black – some powered by sail, others by steam, and a few by both – were substantial in every respect. There was no nonsense about serving up 'foreign foods', and the sofas were as snuff-coloured and as heavily-buttoned, and the saloons as decorated with potted palms as they were at home. Moreover, because almost all the coaling and cabling stations were British in character as well as in fact, in the oddest places there were British insurance brokers, ship's chandlers, surveyors, marine engineers and victuallers. The Royal Navy dominated and protected the sea routes, and so in the great shipping lanes and ports there were generally British warships: the paintwork a glittering white in tropical waters, elsewhere black with orange funnels and white upperworks; the decks alive with barefooted seamen; the bridge staffed by officers as impeccably turned out as the pipe-clayed rigging, the polished handrails, the cheeseddown cables and the gleaming brasswork.

Arriving at Penang or Singapore, Manila or Hong Kong the plant hunter would have to change ship to a tramp or coaster; that is, from sparkle, spit and polish to coaldust, grease and scruffiness, the sofas and potted palms giving place to cane planters' chairs and canvas punkahs which stirred the foetid air. The food, though, would still be stodgy and more suited to the German Ocean than the South China Sea because a large percentage of small merchant shipping sailed under the Red Ensign and kept to English ways. But then, put ashore at Soerabaja or Geelvink Bay, at Fakfak,

Potianak or Yap, he instantly lost touch with the West and found himself on the edge of a wild and alarming experience.

That many collectors persevered and penetrated to the interior of Malaya, Indonesia, the Philippines and New Guinea and their outlying archipelagos was a tribute either to their vitality and endurance or to the strength of their conviction that the plants of the tropical East Indies were well worth the ghastly risks they ran through the climate and terrain and at the hands of head-shrinkers, poison-dart blowers, ferocious beasts and venomous reptiles. The two great Orchid grounds, for example, in Java and the Philippines, were full of obstacles: the jungles so densely matted that tracks had to be hacked out, and teeming with huge slugs and leeches, insects and fruit-eating bats. But they were prodigal in beautiful Orchids: species of Aerides, Arachnanthe, Coelogyne, Cypripedium, Dendobium, Phalaenopsis, Platyclinis, Sarcohilus, Spathoglottis and Vanda, and the largest tree-perching Orchid of them all, Grammatophyllum.

So rich were the veins of Orchid gold that professionals were sent out with the sole purpose of finding new species. They were required to note the location of other plants if possible, but their real business was to collect commercially profitable Orchids. Some English nurserymen became internationally famous for specializing in Far Eastern plants: Loddiges of Hackney Wick with a reputation for nobbling missionaries, Hugh Low of Clapham who had the advantage of being father to Rajah Sir Hugh Low of Sarawak, Standish and Noble of Bagshot, and, of course, Veitch. They competed with one another in the salaries and bonuses they offered, and because these were often very large Orchid hunting attracted adventurers as well as respectable professionals. Indeed, it seems that some of the former were as ruthless in their pursuit of spoil as the pearlers and blackbirders who enriched themselves at the expense of everyone else and who made the Coral Sea between Thursday Island and the Solomons a living hell for countless divers.

William Lobb's brother, Thomas, was the leading Veitch collector in the East Indies and undoubtedly their greatest Orchid hunter. He sent back other exotics, including a collection of tender Rhododendrons, but by inclination he was an Orchid hunter. Unlike William, who – so far as conditions in South and Central America and the Sierra Nevada allowed – was inclined to be a socialite, Thomas was uncommunicative and when a word could be dragged from his oyster-like lips it was almost always about Orchids. He paid for his devotion with a leg – lost on an Orchid hunt in the Philippines – but lived to retire to his native Cornwall. There he enjoyed a few years of retirement at Devoran, saying next to nothing and thinking about nothing at all but Orchids. In the church there is a memorial to the brothers:

> Two collectors of plants from foreign countries, who rendered distinguished service to British horticulture.

And their memory is literally kept green in the graveyard where four specimens of their introductions flower each summer in the balmy Gulf Stream climate: a rose-coloured Escallonia, a crimson Tricuspidaria, a yellow St John's Wort, and an orange Barberry.

Charles Curtis was another of Veitch's Orchid hunters who gave his name to the Lady's Slipper, *Cypripedium curtisii*. Here was an Orchid with a history almost as mysterious as the lost Cattleya which turned up as the corsage of a lady at a Paris ball. Not that its finding was in any way romantic, as Curtis simply saw it growing in the Penang Botanic Gardens and sent it back to England. It was much admired for its violet clog-shaped lip and leopard-spotted petals, but either being fêted did not suit it or it pined for the climate of Penang. Whatever the reason, in time it died, and orchidists gradually came to the alarming conclusion that it had been unique. Nothing was known of its jungle home. Curtis by this time was hunting the flora of celestial fields, and the authorities at Penang Botanic Gardens confessed they were ignorant of the Orchid's origin. On the theory that most probably it had been found in the Malayan uplands, other plant hunters in that area kept a constant eye open for it and several went out to the East with the primary purpose of finding it. But no one was ever successful. The Orchid remained as elusive as the Scarlet Pimpernel but Curtis, apparently, had never forgotten its pretty elegance. At the beginning of this century a Swedish plant hunter called Ericson was collecting in the Sumatra Highlands, once one of Curtis's favourite hunting grounds, and being forced by a storm to take shelter in a convenient hut, he found a picture of *C. Curtisii* painted on the wall. Under it was a scrawl:

> 'C. C.'s contribution to the adornment of the house.'

It was nothing more than an elaborate doodle, a time-passer, neatly painted by Curtis years and years before, perhaps to while away the boredom of being trapped, as Ericson was, by a monsoon. But it showed how clear his recollection was of the Orchid which had his name and which had gone perpetually out of circulation.

The repeal of Britain's Glass Tax in 1845 coincided with the flourishing of industries which produced good materials at a low cost. Moreover, the price of coal for heating glasshouses and labour to maintain them was very small indeed. Everything combined to give the designers of glasshouses their head, and children at their first pantomime could hardly have been more excited. They raised what the Poet Laureate of the day eulogized as 'Tropic Squares', which varied in size and grandeur from backyard lean-tos for the culture of Calceolarias to Decimus Burton's Palm House at Kew, a graceful swan of the glasshouse world with 45,000 square feet of glass tinged to a Stourbridge green. The United States of the time, still too busy internally to be particularly inventive, was content to select and copy the best Europe offered in the way of glasshouse designs. Massive stoves and temperate houses, cold houses and conservatories mushroomed in proportion to American industrial expansion, and with a good many other things the greenhouse followed the railway west.

Filling all these glasshouses created an unprecedented demand for exotics, and the thin stream of plants sent from the Eastern Tropics in the pre-Wardian era swelled to a sizeable river. Hunters scoured the Malay Peninsula, Java, Borneo, the Moluccas and Celebes, Guinea and the Philippines. The ubiquitous Orchid was the most profitable quarry but the swamps and forests held many other treasures. Members of the Dracaena family were very popular, and named after such British military worthies as Sir Garnet Wolseley and Lord Roberts. Then there was the Cissus genus, which gave the Californian patio its Evergreen Grape, *C. capensis*, and the Victorian conservatory a climber, *C. discolor*, with light green leaves and flowers the indeterminate colour of opals. And there were Grevilleas, Anoetochilus, scarlet Glory Trees – some deciduous and others evergreen – Ixorias (called 'prince of stove-plants'), and Crotons with foliage as gorgeously and diversely coloured as trigger-fish.

These beauties, and many more, reigned in the crystal palaces of Europe and the Americas until the cost of fabric maintenance and fuel and labour eventually made extensive glass-gardening solely the private province of the very rich and the public province of national institutions.

Reigning with the beauties were plants which intrigued because the Victorians dearly loved curiosities. The East Indies was a centre of the sort of weird vegetables they liked: Swamp Palms, for instance, and Mangroves, massive Bamboos and Lilies. Then there was the Liana which grew to a huge size and stifled the forest trees through which it writhed, or – and this was a delicious absurdity – sometimes it would strangle itself, offshoots and tendrils tightening like fingers round its own throat until it had committed suicide. Still more grotesque was the Sumatran Giant Arum, *Amorphophallus titanum*, a gross tropical Parson in the Pulpit. Whenever the monster flowered in captivity it caused a sensation equal to that of the dramatic sex life of pandas. Not without reason. The parson part is as tall as a real parson. The pulpit is bell-shaped, frilly-edged and has a mauve interior. The whole thing stinks. The flower which bloomed in the New York Botanic Garden in 1937 was said to 'reek of the effluvia of drains and all corruption'. Another flowered in the Leyden Botanic Garden in 1954 and drew crowds of Hollanders eager to savour its 'stench of rotten fish and burnt sugar'. But even more fascinating, and given pride of place in Victorian collections of curiosities, were the carnivorous Pitcher Plants native to the Eastern Tropics.

Linnaeus, with his sense of the appropriate, had named the plants *Nepenthes* after the cocktail made of an Egyptian drug and wine which Homer described as banishing grief and rage and inducing forgetfulness. To a botanist, explained Linnaeus, the effect of discovering a Pitcher Plant would be precisely the same. He was entirely correct.

The most bizarre feature of that bizarre plant is that the pitchers are not flowers but extensions of leaf midribs. Attached at first by a short thread to the leaf tip the pitcher begins as a pale-green disc the size of a shirt button, swells, grows in length, and becomes a jug-shaped vessel with a hinged lid and a lip. It is obviously designed to dazzle with luminous, iridescent colours. And because small honey glands freckle the inside of the jug it promises sweetness. In reality it is a foolhardy trap worthy of Dr Fu-man-chu's most devilish ingenuity. What appears to be an elaborate banqueting hall for insects is their prison and their grave.

The pitcher's lid is involuted, lobster-pot fashion, designed to keep the victims in, and the inside is slippery. An insect attracted by the colour and lulled by the sweetness abruptly finds itself sliding helplessly down into a pool of liquid at the bottom of the jug. If it has the spunk to try to crawl away, it meets downwards-pointing hairs, fouls its legs and wings, exhausts itself and topples down a second time to drown in the stinking mess of digestive juices and water made salty by decomposing corpses. In time, like a body in lime or acid, it disintegrates and leaves few remains.

This carefully-contrived apparatus is not a divine whimsy. It has a serious function which falls well within the ordered pattern of creation because being neither an epiphyte nor with roots in peaty swamps the Pitcher Plant needs nitrates. Corrupting crickets, flies and beetles, spiders, ants and butterflies make up for its lack of nitrogen. Moreover, the Pitcher Plant is not the master of all insects. Two parasites use the plant. One is a fly which as a larva wriggles into the charnel house to gorge on disintegrating bits of insect; the other is a moth which calmly lays eggs inside the pitchers and these, when hatched as larvae, spin silken aerial ropeways and bridges across the jar so that they will not slip into the death pool below, and then eat away the vegetable walls surrounding them until their host is entirely destroyed.

Rajah Sir Hugh Low had first discovered the Pitcher Plant in 1851 but, to the mortification of his father the nurseryman, he failed to introduce it to the West. The silent, Orchid-struck Thomas Lobb did manage it. Nepenthes grew for the first time in Europe but either the stock was feeble or mistakes were made in looking after it, for it faded and died. Veitch wasted no time in wringing his hands. As hothouse evergreens Pitcher Plants had a future. They shared with Orchids what was to him the advantage of being rare and costly to maintain. He judged that in Victorian society they would become a much desired caste-mark, and that Victorian society, having too much time on its hands and being bored, would welcome a novelty. Further there was something about Pitcher Plants which seemed to slot in with the contemporary cult of decadence and esotery. The stuffiness of the time was barely more than a thick coat of paint which hid some surprising realities: that the Queen smoked, for instance, and discreetly took whisky in lieu of tea when out with the ghillie Brown, and that, with many high ranking and intelligent people in the Old World and the New, she took an abnormal interest in spiritualism, mumbo-jumbo and the occult. Veitch judged well. The Pitcher Plant was to become the most delicious hothouse flesh-creeper of them all, and be honoured as a horror by Joris Karl Huysmans an apostle of decadence in French literature:

> . . . from the tip of each dark, long leaf there hung a green string – an umbilical cord – supporting a greenish urn whose interior was covered with hair.

The description, sensational rather than accurate, mirrored the plant's appeal which Veitch was determined to take advantage of.

F. W. Burbidge, a plant hunter with exceptional powers of endurance, and with resource and persistence to match, was sent out to the Far East with the sole purpose of collecting Pitcher Plants. Nothing else. It was 1877. In Delhi Queen Victoria was publicly proclaimed Empress of India. Czar Alexander II undertook to 'chastise the treacherous Turk' and marched on Constantinople. Veitch's ironside collector landed in Borneo.

The first Pitcher Plant had been discovered in Sarawak on the north-west coast. Burbidge estimated, correctly as it happened, that Borneo was the heart of the Nepenthes country, and that somewhere in the shady, sodden forests between Banjermasin and Balakbar he would find what he was after. He did. Shining through the gloomy jungle like cottage lights on a winter's evening he found Nepenthes with colossal, brightly coloured pitchers: one species a deep violet with gaping orange mouth; another in toadstool array – green-spotted with blobs of crimson; another green spotted brown; a particularly large species – later named *N. sanguinea* – the colour of raw beef; and the *albomarginata* species, a ghostly combination of green and white.

Burbidge's cargo was sensational, and all Veitch's resources and skills were used to ensure that like haughty brides they were maintained in the state to which they were accustomed. Within seven years forty-seven species had been established in European and American stoves. It became evident that Linnaeus's name was particularly appropriate. Not only did those who hunted the Pitchers experience the sort of happy forgetfulness induced by Homer's *nepenthes*. Their cultivators, too, were daily comforted as they syringed and stoked to maintain East Indian conditions in the stove. And, in the end-of-winter excitement of re-basketing their precious Pitchers, they were as entirely oblivious of the painful past as any hero of the *Odyssey*.

4

The Land of the Rising Sun

Japan's impenetrability continued long after Dr Ward's triumphant experiment in 1834, and botanists, made greedy for her flora by the reports of Kaempfer, Thunberg and Siebold, pined fruitlessly for the opportunity to go there. She was an enigma, and she remained so until 1859.

It was known in the West that though theoretically she was under the rule of a Mikado Japan had been ruled by hereditary Shoguns or overlords since 1603, and it was the Shogun's irritating policy to ban contact with other countries and cultures. But it was not known how beneficial this had been to the country: that although there had been occasional religious upheavals, agrarian riots and squabbles between bored warlords, old Nippon, previously so bellicose, enjoyed 256 years of peace and great progress. Inevitably the West's knowledge of the country was limited and often distorted, and, as often happens, fable stepped in to fatten history. In the process of time the four main islands and their archipelagos were invested with romance and glamour, heightened by the fact that from the sea forbidden Japan looked strikingly different from any other landfall.

This was not an illusion.

Westerners were to discover later that though as a nation Japan was ancient and possessed of a long and uninterrupted history, geologically she was an infant. The turbulent upheavals which formed her structure had not been honed down by time and rounded off. She was unweathered, craggy with steep mountain-sides and sharp drops where the land had faulted. Most remarkable of all, the earth itself was mutable. Nowhere in the world was land so fevered and shaken with regular quakes and tremors, so threatened with tidal waves. Nowhere was it so thin a crust over the interior sub-terranean fires. Volcanoes erupted. Hot springs gushed from the mountain slopes. Hundreds of pools of boiling mud gurgled and then vomited up into the sky. These pools were deep, apparently stretching farther down to hell than the spires of Western cathe-drals stretched up to heaven. Fires were so common that the few Dutch traders on Deshima who attempted the formidable task of mastering Japanese discovered there was a complete language of fires to describe the range of possibilities from a small and easily dowsed domestic flare-up to a city-consuming holocaust. Their reports of such phenomena seen with their own eyes, were hardly believed. But the Western world was fascinated.

Various aspects of historical interpretation account for the over-throw of the Shogunate and the opening of Japan.

Down in Yedo (later to be called Tokyo) the Shogun ruled as overlord. Up in the old capital, Kyoto, imperial courtiers wore their ceremonial silken clothes, raced crickets in jars, read the sentimental novels of Kioden and Bakin, wrote poems restricted to thirty-one syllables in praise of love of wine or the beauties of nature, gathered together to view the flowering of a Cherry or a Maple in autumn leaf, organized 'moon-gazing parties', composed music for the koto, the samisen, the kokin and the sho, and drank quantities of tea. For far too long they were about as representative of the vitality and will of the Japanese people as the extremely Brittanic Mikado and Pooh-Bah, Ko-Ko and Nanki-Poo of Gilbert's 'Titipu'. But because their complicated love affairs were so spiced with intrigue they finally learnt the fascination of plotting *qua* plotting. And the government down in Yedo was beginning to wear badly. It had many of the faults of old age – shortsightedness, irritability, indiscretion, overconfidence, and a mistrust of change, and no doubt, even after so long a period of virtual peace and much prosperity, there were real causes of discontent.

Yet an equally powerful cause for the opening up of Japan was the peculiar Western determination to 'improve' people. The impertinence of the Old and New Worlds in presuming to interfere in the affairs of so civilized a nation as the Japanese was almost breathtaking; but in 1853 it appeared to be quite natural for a President of the United States to send a letter to the Shogun suggesting a commercial treaty – and send it by four black warships under the command of Commodore Perry. The Shogun was a boy.

His regent temporized, hoping that old Nippon's privacy could be maintained. Then Russia sent a warship to Yedo Bay on a similar errand; and Commander Perry turned up with a second and far more insistent presidential suggestion, and twice the number of intimidating warships.

With great reluctance certain treaty ports were opened to the West and an extra-territorial concession was granted to resident foreigners which gave them the right to be tried under the law of their own nation and by their own countrymen. As a face-saver the Shogun's regent publicly expressed his opinion of the 'red barbarians' and 'foreign devils' who had forced their gunboat policies on to Japan. But the national humiliation was his downfall. The Shogunate was discredited, the regent murdered. To their dismay the plotting courtiers of the Mikado had to leave their sybaritic life in Kyoto and take over the government in Yedo. But nothing now could stop the West. Piece by piece the country was opened up to foreigners.

One of the first to take advantage of the new concessions was Philipp von Siebold, the collector whose ophthalmic skill had given him such high credit with the Japanese until he bribed the Imperial Ambassador to part with a map of Nippon.

Since 1830, when he returned to Holland with a total of 458 Japanese plants, Siebold had engaged more in polemics than either medicine or botany. If commissioned portraits ever indicate the truth, the painting of Siebold, 'a fierce, beetling-browed, bearded face surmounting a challenging figure in uniform', might account for this. He had had cause to be vexed. At Antwerp he offered his plants to the authorities and for some unknown reason they had hesitated to accept them. He instantly removed the collection to Ghent, where it was cared for in the Botanic Gardens. At this point the revolution of Walloons against the Dutch broke out. It was a stiff contest between Catholics and Protestants, liberalism and autocracy; but the issue was decided at a distance by France and England who, between them, set up the independent sovereign state of Belgium with Queen Victoria's Uncle Leopold as king. As a Bavarian Siebold took small interest in the affair, but on Deshima he had been an employee of the Dutch and he was forced to leave Ghent. His plants there were confiscated and, without any sort of authority, distributed amongst the city's nurserymen.

For nineteen years he quarrelled merrily with government

bureaucrats, Dutch East India Company officers, and fellow physicians. In tranquil moments he enjoyed botany and horticulture and the memory of his unusual experiences in Japan, and he consumed his energy by writing a large, illustrated *Flora Japonica*, a *Fauna Japonica*, and a thick *Catalogus Librorum Japonicorum*. Once the Wardian case had been proved as a means of transporting exotics he established a nursery at Leyden to receive and care for imported trees and shrubs from the Orient. Then in 1859, when traders of the Dutch East India Company considered that the troubles between Shogun and Mikado could be exploited to their own advantage, they invited him to go out to Deshima. They recalled his influence with the Japanese which would be useful to the company.

Philipp von Siebold went, overjoyed to be returning to the land of bubbling fumaroles and volcanic eruptions, where earth tremors were poetically described as a great subterranean fish waking from sleep and wriggling in ecstasy; a land of tea-drinking, bathing ceremonials, cormorant fishing and bribable Imperial Astronomers. He had the great advantage of being remembered by the Japanese who, because they honoured him as an eye-physician, were very content to forget his banishment and set him up in a house on the mainland. This made him comfortably independent of the Dutch Company, and because of the extra-territorial concession guaranteed by the government to resident foreigners he was comfortably independent of the Japanese as well. His obligations to both were merely formal, and perfunctorily carried out. His house grew into a palace. His influence enlarged at the same rate. In considerable state he made long and highly enjoyable botanical expeditions into the primeval forests, where he found a rich variety of trees. There were Cryptomerias and Japanese species of Hornbeam, Hazel, Chestnut, Pine, Cypress, Maple, Alder, Oak, Catalpa, Birch, Beech – one now called Siebold's Beech – and Fir. There was even a Japanese Flowering Ash, a lush variety of *Fraxinus ornus* with autumn leaves a metallic violet colour. The shrubs growing beneath the trees or in clearings were equally diverse. They included a host of Bamboos, almost fifty of them, with euphonious local names like Nahiria-dake, Taimin-chiku, Kan-chiku, Kan-zan-chiku, and Médaké – the species Siebold introduced to Europe and which, for a long time, was the only Bamboo commonly grown there. Another of his introductions, the shrub Helwingia, was a botanical curiosity, for its tiny pea-green flowers appeared to sprout from the middle of the tapering leaves. And then there were Daphnes, Weigelias, Fatsias (one species enjoying the notable name of *Fatsia horrida*), a Japanese Buddleia, Andromeda and Box, Siebold's Barberry, sweet-scented Osmanthus, Acanthopanax, and a member of the Rue family, a Zanthoxylum, called Japan Pepper. To Siebold's surprise he found another member of the Rue family, the Orixa, being used as hedging material. The Japanese liked it because its leaves were aromatic and it gave off puffs of scent when brushed by their loose clothes. They appeared not to mind its habit of distributing seeds like a Balsam over an area of several feet and which, because the seeds travelled at the rate of peas from a pea-shooter, made the hedge quite unapproachable at seeding time.

To go on finding month after month fresh species of trees and shrubs and flowering plants proved to Siebold the unsuspected richness of Japan's flora. Because she had been impenetrable for so long, many of the species of her cultivated plants were even entirely new to the West: the Sawara and Hinoki Cypresses, for instance; the Japanese Arbor-Vitae which later in the century the Veitch hunter Maries did find growing wild in the mountains of the Central Island; and the Amelanchiers which were frequently grown beside the wood and lacquered gateways into shrines and temples.*

The Bavarian botanist lived regally, venerated by the local Japanese and especially those with eye troubles, and respected by the central authorities as a *sensei* or teacher because he was so evidently a man of books and culture and cared for natural beauty. The Dutch, though, took a very different view of their appointed physician-agent who had failed to do what was expected of him. Expatriates invariably arouse suspicion amongst their late compatriots, especially if they appear to be happy. Moreover, Europeans are often vexed when – as the patronizing expression has it – one of their number goes native. Philipp Hans von Siebold appeared to have done just that, and because it was improbable that he would pay much attention to a direct recall from the Dutch East India Company they decided to winkle him out with a trick.

* Because *Amelanchier asiatica* has been found as a 'native' of Hupeh province in China it is called the Chinese Service-Berry, though W. J. Bean noted that it was originally introduced there from Japan.

Perhaps his great good fortune in being a totally free plant hunter in a land with so large and interesting a flora reduced Siebold's wariness, or maybe his feeling of well-being made him look kindly and trustingly on men whom he should have remembered were fairly ruthless entrepreneurs uninterested in botany without guilders. Shrewdness, anyway, has never been the hall-mark of impetuous men. Whatever the Dutch traders promised him was simply a bait which he swallowed entirely and he left the security of his Japanese palace and returned to Holland. There, finding he had been duped, he refused to have anything more to do with the Dutch. His collections of oriental trees and shrubs at Leyden were secured by Simon-Louis and taken to France, and Siebold himself, unable to face the business of trying to set up again in Japan, went off to his native Bavaria to die of ennui and bad temper at Wurzburg in 1866.

Though a new era had begun in 1859 only a plant hunter with the prestige of Philipp von Siebold could collect exactly where he wished. John Veitch, who had moved the family firm from Exeter to London, went to Japan himself in 1860 but, being unknown there, his collecting was severely restricted. He was bound to keep within a certain distance of the treaty ports and his movements were under constant surveillance. A vainer man would have found the impositions intolerable; a less cautious one would have paid no attention to the regulations. But Veitch was neither. Like the first Dutchmen on Deshima who, for the sake of trade, had consented to humiliate themselves and spit on the Cross, he saw the value to his firm of submitting to Japanese caprice and he kept to the rules. His plant hunting, therefore, was confined to a small area, and most of his collections were made in Japanese gardens and nurseries, but the enterprise was still highly profitable in terms of commerce. He was the first to introduce the Japanese Fir, the Tiger-tail Spruce, the Japanese White Pine and Black Pine, and Veitch's Silver Fir, a Wedgwood-blue beauty of great value to town gardeners because though a calcifuge, it happily grew in urban surroundings. He also introduced in quantity the Japanese Umbrella Pine,* described by

---

* *Siadopitys verticillata*, the Japanese Umbrella Pine, is so called because the arrangement of the cladodes, or 'needles', closely resembles the ribs of an umbrella. It should not be confused with *Pinus pinea*, the Umbrella Pine of Europe (though this, confusingly, is occasionally listed as Stone Pine), which has a broad, spreading head on a trunk bereft of branches and so looks like an open umbrella.

Thistleton-Dyer (later the third Director of Kew and, as Sir Joseph Hooker's son-in-law, part of the Hooker botanical dynasty) as unique amongst conifers and 'conjectured to have come down to us from a remote geological past which has obliterated all trace of its immediate ancestors'. But the supreme success of John Veitch's short yet productive visit was his introduction of *Lilium auratum pictum*, the Golden-Ray Lily of Japan.

Robert Fortune, a notable collector in the Celestial Empire, was also drawn irresistibly to Japan. Most probably he was curious to see as much as he could of a country emerging from obscurity, but humourless men (and Fortune was one) will never admit to the inoffensive and natural practice of nosy-parkering. And so, in his book *Yedo & Peking*, he laid a false paper-chase trail by claiming to have gone there in search of works of art. They turned out to be of the vegetable kind. No plant hunter of Fortune's status could have ignored the fascination of new species and even genera which he saw being cultivated in the gardens of Japan. He paid two short visits: the first for a week by Nagasaki and for a few weeks more in the district near Yedo and Yokohama; the second in 1861 in the same area. Like John Veitch he was restricted by the Japanese; like Veitch he accepted the restrictions. They were treading where only Siebold had gone before and so much was new and of botanical interest and of commercial value.

Fortune's introductions were of great importance to European gardeners. Hunting out one of them, the magenta-coloured *Primula japonica*, involved no effort at all. 'I shall never forget', he wrote, 'the morning on which a basketful of this charming plant was first brought to my door'. As easy to find, because he collected them all from one garden, was the collection of Chrysanthemums which he introduced: 'Some extraordinary varieties, most peculiar in form and in colouring, and quite distinct from any kinds at present known in Europe.' Two reintroductions were of great importance because the original stock had either failed or had turned valetud-inarian: the double-flowered *Deutzia scabra*, and the autumn-flowering and delightfully fragrant Osmanthus which, although a hybrid garden species of Japanese origin, was named after him *O. fortunei*.

It appears odd that, in proportion, Fortune accomplished so much more than Siebold, the doyen of plant hunters in Japan. But, then, he was essentially a professional gardener with an unrivalled technique in packing and shipping plants even in pre-Wardian days, and Siebold was essentially a naturalist who knew and loved the plants and trees of Japan and whose work, although important, made far less mark upon the gardens of the West. Besides this, they represented the two entirely different types of plant hunter in the nineteenth century, who, in turn, were representative of two different kinds of human being – the accomplishers and the appreciators, those eager to do like Occidental gardeners, and those content to be like their Oriental counterparts. Philipp von Siebold confronted, say, by Fuji made blue at the moment of moonrise would very likely have reflected Japanese-fashion on its astonishing beauty and as a scientist, marvelled at its formation. Robert Fortune, confronted by the same inspiring mountain would most probably have endeavoured to climb it.

Though the collections of Veitch and Fortune were mainly from plant-pedlars, nurseries and the gardens of inns, temples and private houses, and they were both obliged to remain within a restricted area, neither had guarantees for their personal safety. While Veitch was tracking down Pines and Spruces and Fortune was writing enthusiastic notes on Chrysanthemums 'with petals like long thick hairs, of a red colour, but tipped with yellow', the Shogun's well-meaning regent was being butchered not far away at the very gates of his master's castle. In 1861 when England mourned the Prince Consort and the bloodiest civil war in history blew America's peace to smithereens, and Veitch and Fortune were busy packing up their Wardian cases, the Japanese Samurai, indifferent to treaties, were busy chopping foreigners to bits with two-edged swords. A year later the Mikado's control tightened, but not before an Englishman named Richardson had been murdered on the Tokaido road between the old and the new imperial capitals. At this enormous affront to British dignity the Royal Navy bombarded, burnt and completely reduced the port of Kagoshima. Simultaneously at the London Flower Show, where many of Veitch's treasures were displayed, visitors were so enraptured by the Golden-Ray Lily of Japan that the ladies caught their breath and the gentlemen, as though in church, removed their hats. It was, in fact, a long time before the radical revolution was completed in Japan and European plant hunters could collect there in any degree of security.

John Veitch fretted until it was reasonably safe to send out a collector. Then, in 1875, he chose Charles Maries, one of the firm's nursery foremen to go to Japan. Maries knew his botany well. He had been taught as a boy by Joseph Hooker's father-in-law, Professor Henslow. He was also a sound practical gardener, fit, twenty-four years old, and with no emotional entanglements at home to make him fretful and take his concentration from the job in hand. John Veitch found that even with all these qualities Maries 'lacked staying power' – and he was never able to match his own introductions or Fortune's. But what he did produce was profitable. Veitch knew quite well that though the Golden-Ray Lily was the cake of the firm's trade, the common bread and butter would matter as much and maybe more in the end. With this in mind he told his collector to send home seeds from as many different conifers as he could. Maries did and from this alone he justified Veitch's investment in him, but he also made other money-spinning introductions. For three years he was in the Far East and for most of it in Japan. Neither of his two trips to China was happy. On the first he found the *gloriosoides* variety of *Lilium speciosum* beside the Yangtze, but even this triumph hardly made up for the appalling attack of sunstroke he contracted there and which drove him in a madness of pain down to the coast. On the second trip he collected seeds of *Primula obconica*, and again the triumph was spoiled for him when he was caught, robbed and beaten up by bandits. He was happier working in Japan, and was the first plant hunter to collect in the northern island of Hokkaido. There, though he found the climate stern in the winter when bitter winds blew in from the Sea of Okhotsk, there were unquenched fires beneath the frozen ground. For all that the countryside looked like Scotland, it was enlivened by the constant threat of earthquakes, volcanic eruptions, gales and tidal waves. It was there that he found so many of the conifers Veitch had asked for, and in addition the Chinese Bellflower, *Platycodon mariesii*, and *Lilium splendens*,* a Lily with dramatic trumpets of light red with livid spots. It was Maries, too, who introduced the Witch Hazel, a shrub which was to become such a universal favourite that gardeners crowned it as one of the quintet of winter gardens and which, with the ubiquitous Daphne, Winter Jasmine, Forsythia and Japonica, reigns to this day. Unlike the majority of Veitch's plant hunters who retired to

* Its synonym is *L. thunbergianum sanguineum.*

dream of their foreign travels, Maries showed an original streak in settling in India as horticultural adviser to an Indian prince.

The foundation in 1872 of the Arnold Arboretum of Harvard University and the work done by its director, the sylviculturalist Charles Sprague Sargent, had a large effect upon the collecting of trees. He specialized in the trees of North America, but he also collected in Japan. He was there not long after Maries, his visit coinciding with that of the most spectacular of all plant hunters in Japan, Père Urbain Faurie.

Faurie was essentially a botanist and far less interested in the shanghaiing of plants from their native habitats than in identifying and preserving them. He went native in a way that would have warmed Siebold's heart, taking the trouble to learn what he could of Japanese and generally dressing in the slack and comfortable clothes of a coolie. No foreign botanist had such an intimate knowledge of the flora of Japan; nor, apparently, such energy. Faurie was said to have scaled 90 per cent of Japan's hills and mountains, and he was quite happy to get up twice or three times during the night to change the drying papers in his botanical presses. In twenty-four years he collected 22,500 specimens – all carefully dried, preserved, identified, named, labelled and filed – and his influence on local enthusiasts was very great. Far earlier than China, Japan began to produce her own expert plant hunters and send specimens of what they found to botanical gardens in the West. The botanist Shirasawa, for example, discovered the Japanese Douglas Fir in 1893 and was responsible for its introduction to Europe and the United States. With their aid, and the careful concern of 'Chinese' Wilson, who was sent there to collect by the Arnold Arboretum during the First World War, the flora of Japan was systematically catalogued and analysed. By any standards it was impressive and present-day Western gardens without the plantings originally native to the Land of the Rising Sun would appear strikingly incomplete.

Post-Wardian China

A commercial war and the introduction of Wardian cases entirely revolutionized the export of plants from China. Britain's role in what came to be known as the Opium War was fairly indefensible. The East India Company's monopoly ended in 1834, but the British government, determined to keep an eye on Chinese trade and affairs, sent out a representative or Plenipotentiary Minister to act on its behalf. He was the agent of British gunboat policy rather than an ambassador accredited to the Imperial Court.

The Manchu Emperor Tao-Kwang was favourably disposed to the Europeans and Americans who traded from Canton and Macao. His predecessors on the Dragon Throne, indifferent to Western sophistication, had aloofly called Europe 'the Western Ocean' and Europeans either 'Franks' or 'Red Hairs', but Tao-Kwang was intrigued by the culture, the guns and the astronomy of the Westerners who paid him court at Peking. Yet, though favourably disposed and intrigued, he was not prepared to grant them permission to penetrate into the interior of China. Nor did he hesitate to condemn and forbid their irresponsible introduction of opium as an article of trade. It was having disastrous results. In exchange for the enervating and expensive drug which the British in particular were illegally importing from Bengal and England, his country was being drained of silver, tea, silk, porcelain and other profitable products. He complained of this 'supervised smuggling' to the British Minister.

Then as now dope peddling was a lucrative business. The Minister paid no attention. To his surprise an imperial commissioner named Lin arrived in Canton with great powers and explicit instructions to put a stop to the smuggling at any cost.

Lin happened to be a man of integrity and one of the few imperial servants who did not have a hand in the opium trade. He served his master well, showed he was not open to British persuasion and, to everyone's dismay, he stopped all foreign trade, imprisoned foreign merchants and employees in their factories and presented them with an ultimatum. His sanctions would continue, and for ever if necessary, unless all the British-owned opium lying in the warehouses and in store-ships at river moorings was given up. The Minister tried every method to make him change his mind, but Lin was inflexible. The British had to capitulate and no less than 30,000 fair-sized chests packed to the brim with opium were surrendered to the commissioner. It represented a fortune, and Lin destroyed it. But he was slow to concede old trading rights. There were tales, too, of atrocities committed on prisoners in the trading factories. Claiming that Lin had not kept his part of the bargain, the British government declared war.

A fleet and an army arrived in 1840. The island of Chusan was seized, Canton, Ningpo and Shanghai captured. Warriors and diplomats of the venerable Celestial Empire could not withstand the pressure put on them by professional Western soldiers who, in the days of Victoria, were always spoiling for war. The treaty which followed was of great benefit to Britain, and therefore to Western plant hunters.

By 1842 the ports of Shanghai, Ningpo, Fu-chu and Amoy were open to the trade of the world as well as Canton and Macao. Hong Kong was formally ceded to Britain in perpetuity. The Chinese undertook to be less strict in supervising the activities of aliens in the immediate environment of the treaty ports, and journeys of up to fifty miles distant were winked at, if not officially permitted or encouraged. This dispensation, and four additional ports, gave collectors more than 1,500 square miles of extra territory in which to hunt plants.

The rocky island of Hong Kong covers less than thirty square miles but it contains notably different terrain varying from the gulleys of the north to the bleak, thin lands in the south-east. Its climate encourages growth. Being able to botanize at leisure on British territory, several amateurs as well as professionals set about listing the island's flora, and their collections, sketches and notes were used by Bentham (of Bentham and Hooker fame) when he compiled the *Flora HongKongensis*. It turned out to be astoundingly large: almost 1,000 species of 550 genera. Officers attached to the invading force and to the British army of occupation provided four of these pioneer plant hunters: a medical officer of the Cameronians, two naval surgeons, and John George Champion, a captain in the 95th Regiment of Foot. Bentham was particularly grateful to Captain Champion because he was a systematic botanist not simply a collector. Being scrupulously exact, he drew analytical sketches and wrote his descriptions on the spot – not later in the day when his memory might be clouded with tiredness or over-eating. Charles Wilford,* a Kew collector of the new generation, made a large contribution to Bentham's *Flora*, as did the land surveyor and deer hunter from Texas, Charles Wright, who took to botany and hunted plants with all the dash and energy he had shown in hunting deer. Bentham was in his debt for a large number of carefully dried and pressed Hong Kong plants. His *Flora* owed a great deal as well to Robert Fortune and Henry Hance, both of such importance in the history of plant hunting in China that they merit separate attention.

Fortune's private plant-hunting expedition in Japan has already been noted. It was, in fact, his only private expedition, and was the last he made in the Far East. His reputation had been established long before.

By birth he was of Scottish cottage-folk. By habit he was dour, persevering, as unsmiling as David Douglas. By genius he was an excellent descriptive writer with an eye for a 'gardener's plant' and a real flair for collecting. By nature he was eccentric, direct, brave and naïve. These judgements are speculative, lying within the seemliness of historical probability. His books and numerous articles are straightforward and interesting but curiously impersonal. We see little of Fortune in them, and the diaries and letters he wrote which might have mirrored his personality were destroyed by relatives when he died – a curious procedure, motivated either by caution or censoriousness, but in neither case particularly laudable. Yet however enigmatic Fortune might have been, his place in the history of plant hunting is a high one.

When details of the Opium War treaty became known, it was not long before the Royal Horticultural Society, brooding on those

* When the British Government, anxious to sweeten the Mikado of Japan, sent him the gift of a steam yacht, the *Emperor*, Wilford was added to the ship's complement from the Naval Dockyard out to Hong Kong.

additional 1,500 square miles of territory, appointed a China Committee. Its progenitor was John Reeves, who had retired from his post in Canton and by then lived in London. His knowledge of the Chinese was invaluable, his enthusiasm infectious, his advice that they should pay a collector to go out on the Society's behalf as soon as possible. Fortune was superintendent of the Society's hot-houses at Chiswick and eager for the appointment. He was thirty-one, a hard worker, a skilled gardener, and ambitious. Reeves reckoned he would do, and the Scot was provided with a spade and trowels, a life-preserver, a fowling-piece and pistols, a Chinese dictionary, and long and detailed instruction. His salary was to be £100 a year with an allowance of up to £500 expenses. He must keep a daily journal – to be the Society's property – and write descriptive letters to the Society's secretary: '*This will enable the Society to judge of the progress you are making. All letters must be sent in duplicate, by separate opportunities, so as to guard against accidents.*' When sending plants back in Wardian cases he was '*to impress upon the minds of the Captains the indispensable necessity of the glazed boxes being kept in the light, on the poop if possible*'. He was to understand and remember that '*to all collections of living plants and seeds the Society lays exclusive claim*', and he was to pre-pare for the Society a set of dried specimens of all the plants he came across. He must make inquiries on certain points – why, for example, '*it is a general practice for the Chinese to put up their seeds mixed with burnt bones*', and '*the circumstances under which the Enkianthi grow at Hong-Kong*'. He was required in general to bear in mind that '*the value of the plants diminishes as the heat required to cultivate them is increased. Aquatics, orchidaceae, or plants producing very handsome flowers are the only exception to this rule*'. And he was required in particular to collect analyses of soils with twenty-two other items, including '*the plants that yield tea of different qualities*', '*Peonies with blue flowers, the existence of which is, however doubtful*', '*Cocoons of the Atlas Moth*', '*the Canes of Commerce*', and '*the plant which furnishes Rice Paper*'. A most extraordinary feature of this formidable document was an admission: '*The Society cannot foresee what it may be possible for you to accomplish during your residence in China; which according to their present views they wish to limit to one year*'. It showed that even the experienced John Reeves knew very little for certain about China at that time. As it turned out, Fortune was to travel in the East for over nineteen years and, though his introductions were substantial, he barely took one skimming from the cream of China's flora.

He left England on a ship named the *Emu* and, as the R.H.S. had directed, he messed with the captain. His first sight of the Celestial Empire was disappointing. Like John Gibson approaching the Ganges delta he had expected something very different. As it was, he felt let down:

> Although I had often heard of the bare and unproductive hills of this celebrated country, I certainly was not prepared to find them so barren . . . They had everywhere a scorched appearance . . . The trees are few, and stunted in growth, being perfectly useless for anything but firewood.

Nor did he care for the Chinese:

> From the highest Mandarin down to the meanest beggar they are filled with the most conceited notions of their own importance.

A little experience and time modified his view of both China and the Chinese. The former he grew to love. The latter he regarded half with affection, but half with vexation because naturally they crowded to see him and they were amused by his eccentric use of smoked spectacles, a trap umbrella, and a hat not unlike the tarboosh of a Turk. He showed a certain gaucherie when a Mandarin wanted to make him the gift of a valuable bonsai tree grown in the shape of an animal:

> As it was of no use to me, and my collections of other things were large, I was obliged to decline the present, which he evidently considered of great value, and no doubt wondered at my want of taste.

Quite likely, too, the Mandarin wondered at the Scotsman's lack of affability and courtesy. Foreigner still bit foreigner on the coasts of Cathay.

Though Fortune never went farther than thirty miles from any treaty port on his first expedition, his collections justified all the hopeful expectations of the China Committee. Most of the collecting was done by Chinamen 'engaged at a small daily remuneration', Fortune himself acting as a plant clearing-house. But some plants he found in person, a specially prized bag being the Japanese

Anemone, which wasn't Japanese at all and which he found flowering in November in Chinese cemeteries. On this and later expeditions he also introduced Weigelia, Winter Jasmine, Bleeding Heart, Forsythia, Tree Paeonies and many species of Plums, Primulas, Rhododendrons, Azaleas and Chrysanthemums.

He was by no means immune from physical dangers. On returning to Chusan at the conclusion of a short trip into the interior he lay weak with fever on the deck of a junk. His crew were a timid lot who merely screamed with terror when pirates came into sight and bore down on the junk. Fortune could hardly stand. He waited until the pirates were alongside before raking them with a blast from the heavy double-barrelled fowling-piece provided by the Horticultural Society. They sheered off, considered their chances, and made another attack. Again Fortune blasted as many as he could, and a third time, until they were finally driven off.

With his collections stowed in eighteen Wardian cases on the poop of the Indiaman *John Cooper*, Fortune made the journey back to England in 1845. His courage and his skill were rewarded with the curatorship of the Chelsea Physic Garden but he only held the appointment for two years. The taste of China was in his mouth and when the East India Company asked him to go out to collect China Tea for growing in the highlands of India he accepted at once. The commission was not easy. The Chinese guarded their Tea industry with such jealousy that penetrating to the gardens and collecting seeds involved a great deal of personal risk. His 'cover' as a botanist was a good one and particularly convincing because there were many wild plants in the Anhwei and Bohea Tea districts. And making so bold as to pretend to be a native from a distant province, he managed to carry it off. For three years running he sent supplies of seeds to Calcutta and the second and third consignments, travelling as germinating seedlings in Wardian cases, were particularly successful.

In the potted biographies of Fortune he is often described as 'the introducer of the Tea plant into India' in 1848. Research by botanists, growers, blenders, plant hunters, geographers and explorers suggests that though the Tea Ceremony had been carried on in China for at least 3,000 years, and it has long been considered the home of the Camellia whose tips make tea, very probably the first tea-drinkers lived in the jungle-covered mountains between Assam and North Burma, especially in the region about the lonely mountain called Dapha Bum; and that far from Tea plants travelling from East to West, the contrary is more probably correct. The directors of the East India Company, however, were uninterested in plant ecology. They were principally concerned to congratulate Fortune on his excellent work, because there was a very large chance he would make tea nabobs of them all.

Apart from this achievement Fortune found the Barberry *Berberis bealei*, which 'far surpassed in beauty all the other known species'. Unfortunately there was a belief amongst the Chinese that the plant had magical properties and he found it difficult to overcome their superstitious dread of parting with any specimens at all. The three little plants which at last he wheedled from them were the parents of all the stock now living in Europe and the United States. Another by-product of the Tea expedition was his introduction of the seed of the Funeral Cypress, *Cupressus funebris*. It had been found and described by the botanists attached to Lord Amherst's embassy but not introduced to Europe. Fortune delighted in the tree. Being a calvinist and very much a man of his gloomy country and time, he was fascinated by doom and death and all its trappings, and he immediately decided that the conifer would make a fine ornament in Western cemeteries. John Lindley, by then Professor of Botany at University College, London, agreed wholeheartedly, saying the Cypress was 'perfect for graveyard decoration'. Fortune's seeds were grown in England and several Cypresses were raised, but in the end it proved to be too tender to grow everywhere and would only mature in the freak regions where fingers of England and Wales are warmed by the Gulf Stream.

Fortune's success as a Tea smuggler earned him the East India Company's commission to collect and transport more Tea plants and seeds. He found China seized in the Taiping Rebellion, a conflict between the Manchus and Mandarins and a revolutionary movement led by a former village schoolmaster who went into ecstasies, flirted with American Methodism, declared himself to be the younger brother of Jesus Christ, and thereafter lived in splendour with thirty wives and 100 concubines. The internecine struggle was to last twelve years, and in it the Chinese government was forced to invite the military intervention of Britain and France, frontier lands were lost to Russia in the north and to Moslems in Yunnan and Turkestan, and the Dragon Throne was several times in real danger of being overthrown. Quite unconcerned with all this, Fortune took advantage of the lack of supervision to hunt plants in areas which were generally forbidden, and take what he wanted from the unguarded Tea gardens. His bravery, demonstrated by the way he had dealt with river pirates, was never in doubt; nor was his eccentricity in view of the fact that generally he was armed only with a trap umbrella.

Fortune's fourth trip was under the auspices of the United States Patents Office. This was in 1858. The Taiping Rebellion was still chafing and tearing at the Chinese peoples but Fortune barely seemed to notice it. He was engaged to collect plants and study Chinese horticulture for the American government, and he did just that; no more. It was his last sponsored plant-hunting expedition and it confirmed his standing as a collector.

Fortune had, of course, been lucky to be the first front-ranking plantsman on the scene directly after the Opium War. And he had been the first to make extensive use of Wardian cases. But his large success and his fame were mostly due to the fact that he was an instinctive botanical explorer and a thorough professional in the handling of seeds and plants.

The fame of Henry Hance, an assistant to the consul at Canton and afterwards vice-consul at Whampoa, rested on different foundations. He, too, collaborated in providing material for Bentham's *Flora HongKongensis* and, though his rank in the Consular Service was low, his influence in the field of botanical exploration was large, and widespread in the south of China. Maybe the two facts were related. Hance was an Englishman with an unusual talent for languages. He spoke French and German as well as he spoke English, and if the classical tongues had been living instead of moribund and confined to scholars and churchmen he could have done the same in ancient Greek and Latin. But, though he lived forty-three years in China, and came to love and understand the country as well as any other Englishman of his time, he never managed to learn either Mandarin or Cantonese or even the 'Coolie Talk' which was mangled Mandarin. Because this checked his career in the Service some of his contemporaries believed he was simply being stubborn. They overlooked the fact that it was impossible to study Chinese in the same way as any other language; that the Written Style is undoubtedly the most difficult study in the world; that confronted by five variations of Mandarin, each requiring as much definite study as a separate language, Hance might well have found the task too time-consuming for his taste. Or, if he did not care for responsibility and promotion, lacking Chinese was an easy way to bar his progress. At any rate his junior position allowed him sufficient leisure and freedom from worries to concentrate on the matter closest to his heart, the study of plants; and his remarkable administrative powers, lost to the Consular Service, were used in the service of plant hunters for many years. What Reeves had been in Canton and Wallich in Calcutta, Hance was to be in the humble vice-consulate of Whampoa. If collectors wanted advice or coolies or merely someone to talk shop to they went to him. In return he asked no more than that if possible they should send him duplicates of the plants they found. Only occasionally did he hunt plants himself, but he knew and he helped so many collectors in his forty-three-years' service that his herbarium became larger and larger. It is now stowed at the British Museum (Natural History) in London – a testament to his knowledge of botany and his talent for administration.

Though he was internationally known and respected even Henry Hance could not be at the centre of collecting in the whole of China. The country is too large, stretching over 2,600 miles from Moha in the north to Yaichow in the south; that is, the distance from New York to Bogota or from London to Timbuktu. His sphere of influence was primarily in the south and his herbarium represents the flora of that part of China.

The most influential botanist in the north was undoubtedly the Russian Carl Maximowicz, a man who began with the advantage of having learnt his botany at the knee of the great Bunge, and being appointed Curator of the St Petersburg Herbarium at the age of twenty-four. In 1853 he was sent on his first expedition to the Far East, travelling out to China by a frigate in the Russian Navy and back by the overland route, and during the next thirteen years he made many other botanical explorations, in Japan as well as Manchuria and the northern provinces, showing a daring and enterprise in looking for plants in out-of-the-way places which was unique at the time. His most important work, though, was as a type of repository of plants from the north of China, and, unlike Hance who assembled his huge herbarium of southern plants in Whampoa, Maximowicz settled at home in St Petersburg to collate his own discoveries with those of other people. For years afterwards Russian botanists out in the field would send him duplicates of the plants they found, aware that he was in no sense a rival as he was uninterested in the commercially profitable business of introducing plants into cultivation. His principal purpose was to mass together as much information as possible on the flora of North China. Both he and Hance were in positions of great importance in 1860 when, as a result of the second phase of her war with the West, China was obliged to open her interior to the Treaty Powers – Great Britain, Russia, France and the United States – and plant hunters could at last follow the rivers to the highlands of the West.

It was an unhappy time for China. The Taiping Rebellion continued right through the war with the West. The war itself was a discreditable affair, a mountain made from a molehill by the West chiefly with the idea of squeezing concessions from the Chinese government. The Manchus and Mandarins, aware that the empire was disintegrating, shamed by their own stupidity and the gunboat manners of the West, were further humiliated by the necessity to ask their conquerors' help in putting down the Taiping Rebellion. The 'Ever Victorious Army', which was badly named but brilliantly commanded by a fireball Englishman called Gordon, was to achieve this for them, but China had already lost something she would never regain.

Yet if it was a bad time for Chinese statesmen and revolutionaries there was no time quite like it for botanical explorers. Joseph Hooker's expedition to Sikkim had shown the great promise of the Himalayan Range. His brief plant-hunting trip in Assam where there was warmth and an abundance of rain showed how generative nature could be under those conditions. Plant hunters in Burma, and those in Indo-China and Siam had never been in any doubt that an enormous store of hardy flowering plants and trees lay in the forbidden mountain ranges to the north. The botanists on the China coast and those who with or without leave had collected in the central plain, had longed to go west for the same reason.

Now they could go.

PART FOUR

The Edge of the World

i

The Mecca for Botanical Explorers

ii

A Revolution in Western Gardening

iii

Private Travellers, Consular Officials,
Honoured Commissioners & Military Plant Hunters

iv

Archetype of Missionary-Botanists

v

Some Changes in the Mode of Plant Hunting

vi

Doyen of Modern Collectors

vii

An Economic Botanist

The Mecca for Botanical Explorers

The great collecting district, which for more than a century was to be the most prolific source of flowering plants in all the world, lay at the heart of Asia. Part was in India and part in Burma, more in Tibet, most of all in the three Chinese provinces of Kansu, Szechuan and Yunnan. It was an area drained by the tributaries and main courses of five mighty rivers, the Yangtse-Kiang, the Hwang-ho, the Irrawaddy, the Salween, and the Mekong. The size and extent of their headwaters, cataracts, rapids, currents and vortexes, and the grandeur of the surrounding highlands, set the scale for the whole region. The first Westerners to go there found that the immensity of everything made them feel Lilliputian. They were intimidated by these massive rivers; the extreme variations of climate; the eternal snows; the towering watersheds; the flat, dry plateaux criss-crossed with river-beds in chasms over 1,000 feet below; the screes over which winds howled and screamed for ever.

The Chinese of the plains had never cared for the mountainous west, and with all the superciliousness of metropolitans they considered themselves a large cut above their provincial fellow citizens. In fact, the west was a bogy to them and the threat of banishment to the three provinces was regularly used as a means of keeping children well-behaved much as naughty Western children were once threatened with the police, or rag-and-bone men, or Bonaparte. But the banishments were an unhappy reality, not simply a menace in the nursery. At a time when the Emperor of China was not only an autocrat but also the nation's chief priest and mediator who at each solstice made offerings of silk, roast meats and wine in the Temple of Heaven, it was customary for offending officials and courtiers to be given appointments 'South of the Cloud', i.e. Yunnan, or in 'the Land of the Four Rivers', i.e.

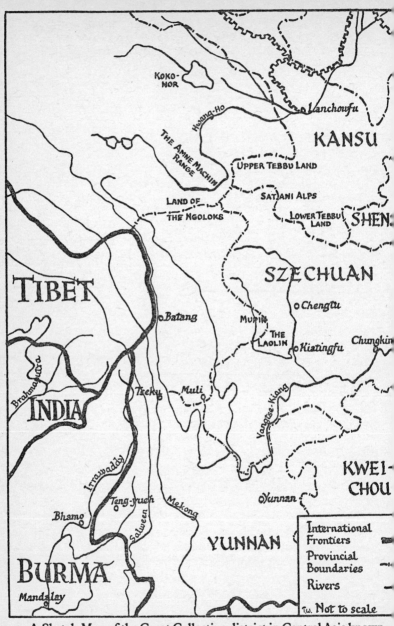

A Sketch Map of the Great Collecting district in Central Asia known as The Edge of the World.

Szechuan. And at the solstice appointments they could even be sent to Kansu, where conditions were so primitive that the only form of transport was by wheelbarrow.* These were much-dreaded banishments, though one exile, the renowned poet Po Chü-i who was sent off to be governor of a remote city in Szechuan, consoled himself by gardening:

> I took money and bought flowering trees
> And planted them out on the bank to the east of the keep,
> I simply bought whatever had most blooms,
> Not caring whether peach, apricot, or plum.
> A hundred fruits, all mixed up together;
> A thousand branches, flowering in due rotation.
> Each has its season coming early or late;
> But to all alike the fertile soil is kind.†

Po Chü-i's pleasure was in cultivating fruit trees for their blossom, but the fertile soil of Szechuan was kind to wildings, too; and after the treaty of 1860 it was to become a Mecca for botanical explorers.

As reports and dried plants came in from the missionaries and officials who were pioneering the west, both Hance and Maximowicz realized that the rumours and legends of the past were certainly true. The newly-opened territory had a flora unequalled by any other region of the world. Moreover, because the climate up in the highlands could be severe the endemic plants were mostly hardy – which was precisely what Western gardeners were to ask for for the best part of a century, supply either causing the demand or vice versa.

It is for this reason that, although front-ranking plant hunters have worked in other parts of the globe during that period, their collections of hardy perennials have never been comparable in quality or quantity with those made in the region which one collector named 'the edge of the world', and so the last part of the study is concerned with some of the plant hunters who have collected there. They were legion and moreover so diverse in nationality, personality, character, approach, collecting methods and degrees of success that the account is necessarily limited and cannot claim to be comprehensive.

* It is still the main form of transport today.
† The translation is Arthur Waley's.

A Revolution in Western Gardening

The members of the China Committee had shown themselves conspicuously in advance of their time in asking Fortune to search for hardy and half-hardy plants; though, being botanists as well as horticulturalists, they would quite naturally be anxious to secure all types of plants and not merely the commercially viable tropicals which were the chief interest of nurserymen. Bigger and brighter exotic flowers and outlandish curiosities continued to hold the attention and the affections of rich men until the soaring price of glass, fuel and labour drove Croesan gardening out-of-doors again; but the Royal Horticultural Society was anticipating the new gardening public which came into existence in the late Victorian age and which had a purse that could only afford the culture of hardy perennials. Their prophet was an irascible Irishman, William Robinson, and he, with the aid of Miss Gertrude Jekyll, was to revolutionize the style of Western gardening.

Robinson hated glass – most probably because as a spiteful gardener's boy who could not bear to be rebuked he had revenged himself on a censorious employer by allowing frost to destroy a handsome collection of stove plants. This led to his permanent exile from Ireland and a loathing for anything in the least artificial in gardening. The curious plants and horror-strikers which gave his contemporaries such thrills of excitement, the heavily-scented, iridescent tropicals and the whole regiment of marvellous orchids which they adored, merely made him mad. So did the dense and shady shrubberies of the period, the knick-knackery, the wriggly paths, the bedding-out schemes. Strangely enough, though he had a passion for plants he abominated botany and was shrill in his condemnation:

The subordination of the garden to Botany has been fruitful of the greatest evil to artistic gardening.

As the apostle of wild gardening and, later, of the herbaceous border – 'a child of sun and colour-drenched impressionism' – he was to tyrannize the gardeners of two continents and popularize the sort of hardy and half-hardy trees and shrubs and perennial plants which were to be found in abundance and in great variety in the 'furrowed folds beneath the hoary heads of the great mountains of western China'. As a result there was the equivalent of a gold rush to that area, a plant rush which brought enormous satisfaction to amateur botanical collectors, equally enormous profits to the professionals and their sponsors, and thousand upon thousand of new genera and species to the gardens of the Western world.

Private Travellers, Consular Officials, Honoured Commissioners and Military Plant Hunters

The first plant hunters in the newly opened region were amateurs; that is, in the sense that they were not paid to collect.

A few were private travellers. Of the highest *ton* was the explorer Prince Henri d'Orléans, grandson of Louis Philippe, who won fame by exploring and describing a huge and unknown area in Central Asia. He was the first to discover a number of favourite garden plants: *Meconopsis chelidonifolia, Rhododendron yanthimum, Incarvillea grandiflora,* and *Primula vittata*. And another private traveller was the rich, dilettantish botanist Antwerp Pratt, an Englishman with a grand manner, a patient temperament, and a preference for being slow and methodical. Primarily he was a naturalist, his especial interests being zoology and entomology, but an unknown plant delighted him quite as much as an unknown moth or fruit-eating bat. His first large expedition was done in style and without a trace of the austerity practised by some hunters who appeared to consider discomfort as an essential feature in the practice of their profession. Pratt took his wife and family with him to China, plus furniture, English luxuries, and every conceivable necessity stowed in crates, brass-bound cabin-trunks, and huge portmanteaux. His ultimate destination was the provinces of Yunnan and Szechuan. Like the corpse of ancient Troy, untouched until Schleimann had the imagination to disinter it, this treasure-store of temperate flowering plants was hardly touched until he had the imagination to believe it might be worth a special visit. He bought a houseboat on the Yangtze-Kiang, appointed a German, Kricheldorf, as assistant collector, and set off upstream. He was a great socialite and enjoyed the company of other European naturalists and explorers on his route into the highlands of Western

China. For this reason his pace was leisurely and his collections were not large. But Pratt's main introductions into cultivation were proved and well-loved in their time, and, establishing the enormous potential of Szechuan, he provoked other plant hunters to follow in his footsteps.

There were also consular officials with a liking for botany. Augustus Margary was one, a brilliant linguist, and a somewhat eccentric young man of great promise. Collecting plants was very much his secondary occupation but ultimately one of his chief interests. He was the first Englishman to travel from the Yangtze-Kiang to the Irrawaddy. It was a secret mission and formidable because it involved crossing over a thousand miles of totally unknown country. For Margary it was purgatory. He was only twenty-six but all through the journey he suffered simultaneously from toothache, rheumatism, pleurisy and dysentery. And when he had completed his mission and travelled a little beyond Bhamo, he was abruptly and inexplicably murdered. Though a mission was sent to investigate his death, the reason for it has never been satisfactorily explained. One view, expressed later, is that Margary had permitted a dog to be carried in his curtained litter. Such a gross insult – which almost amounted to sacrilege – might easily have caused the litter bearers to murder him. Anyone so rash as to tempt fortune in this manner must have been out of the ordinary. Margary was. Whenever he felt lonely and longed for home, he would stand outside his tent, sing 'Clementine', 'Polly-wolly-doodle', 'Michael Finnigin', and 'The Lass of Richmond Hill' and, to the astonishment of his bearers, finish the concert with a thunderous rendering of 'God Save the Queen.'

Chungking on the Yangtze, though hundreds of miles from the sea and roughly in the centre of Szechuan, was a Treaty Port and the British kept a consul there. By accident or design three of the consuls in succession happened to be botanists: E. H. Parker, Alexander Hosie, and F. S. A. Bourne. The last should be honoured by gourmets for his detailed work on Chinese vegetables, while Hosie is best remembered for his observations on the whitewax industry. This he investigated as an enthusiastic expert in economic plants, and found that as a production process it appeared to belong to the fantasy world of *Alice in Wonderland*. A species of insect, *Coccus pela*, was bred by the million on Privets down in the south-west corner of Szechuan, packed in paper parcels of a pound in weight and carried by more than 10,000 coolies (sixty parcels of living insects per coolie) for over 200 miles through some of the most difficult country in the world to the plains, where there were plantations of pollarded Chinese Ash trees (*Fraxinus chinensis*). There the insects were released on to the trees, and within 100 days a layer of wax, a quarter of an inch thick, would be deposited, all over the Ash foliage. The branches were then lopped off and the wax removed by heat.

The newly-formed Imperial Chinese Maritime Customs Service, staffed mostly by Europeans, also produced 'honoured commissioners' who hunted plants in the west with great determination. Perhaps the most famous of them was Augustine Henry*, though by inclination he was not a botanist at all. In fact he had been stationed in Ichang by the famous Yangtze Gorges for three whole years before he began to collect plants, and at first his collecting was sporadic and designed simply to kill time and ennui:

> Oh, if you knew the weariness of an exile's life. I have become a great collector of plants and after exhausting the neighborhood I thought of going into the mountains.

The boredom which drove some exiles to drink had driven Henry to botany. It was hardly less habit-forming and very soon he was a compulsive collector. Because his duties prevented him from going up into the mountains unless he was on long leave he hired and trained Chinese boys to go plant hunting for him. Transferred down to the coast, thence by good luck up into Yunnan, he made an enormous herbarium of over 5,000 different species. By the time he retired from the service the once-reluctant plant hunter was a leading authority on the flora of the Chinese highlands, sufficiently informed to collaborate with Elwes in *The Trees of Great Britain and Ireland* and lecture as Professor of Forestry at Dublin for thirteen years.

With the private travellers and explorers, the consular officials and 'honoured commissioners', there were a number of highly successful military plant hunters. Without exception they were Russian.

Imperial Russia was one of the Treaty Powers of 1860 and she

* He should not be confused with the American missionary B. C. Henry who was a collector of sub-tropical and tropical plants in the southern provinces of China, especially Kwantung.

made use of the privileges granted by the Chinese to mount expeditions of exploration in Central Asia and Mongolia. They were not entirely scientific. Russia was exceedingly fidgety about British rule in India and she wanted information about the country which lay to the north of India. Her para-military expeditions along the frontier were interpreted by the British as a sly attempt to steal away the territories only recently stolen by Lord Dalhousie, 'the Annexer'. Queen Victoria headed her subjects in expressing alarm and indignation at the Tsar's effrontery. But poor Alexander II, though of a moribund dynasty and destined to be blown to bloody fragments by a conspiratorial society named the Will of the People, happened to be one of the few intelligent Romanovs, and he was unmoved by the bogus moral outrage of the Widow of Windsor. The expeditions continued. Their personnel often included a trained naturalist who was also an officer in the Imperial Army. And sometimes these naturalists led their own expeditions with the active encouragement of the Tsar's government. If they were of sufficient standing, large grants were made towards their expenses, and scientific success was rewarded by military promotion, though it often happened that the naturalist had to be content with the promotion and pay for the expenses himself. There was the case of Nicolai Mikhailovich Przewalski, for instance, whose achievements took him literally from being a private soldier to a major-generalship, but who often had to stop and retrace his steps on expeditions because of lack of funds. He was doyen of the Russian collectors in the post-1860 period, distinguished quasi-military plant hunters such as Potanin, Berezovski, Kashkarov and Roborovsi who sent duplicates of all they found to the industrious Maximowicz in St Petersburg, and like the others he was essentially a naturalist, not a specialist in botany. In fact his main interest lay in the direction of zoology and, bearing this in mind and remembering, too, that he travelled rapidly and preferably in wintertime, and that most of his collecting was done in an area far less rich in plants than the floriferous south, it is astounding how large his herbarium became. His great aim was to reach Lhasa, and he died of typhoid on the last attempt in 1888, but before then he had collected over 15,000 specimens which included over 1,700 species. Holy Mother Russia was served well by Major-General Przewalski and his colleagues.

Archetype of Missionary-Botanists

A t various times in the history of China Roman Catholic missionaries had been granted special privileges, but not until 1860 were they able to proselytize in the mountain provinces of the west. In so rugged a region of weathered, tradition-steeped hilltop people their task was formidable and it is not to deny their zeal as priests to say that in the main they were far more successful in collecting plants than in collecting souls. The first of them, Jean Pierre Armand David, was a naturalist *par excellence*; a man of energy and modesty and unfailing amicability, a devoted priest and a candid realist. He soon had the measure of the Chinese hillpeople and he came to love them dearly, but he was not blind to the difficulties. After years of experience he wrote to his godfather:

> Above all do not think China will become Catholic. At the pace things are going now, it will take 40 to 50 thousand years before the whole Empire will become Christian.

Accepting this unwelcome fact in no way lessened David's attempt to convert the Chinese and prove himself wrong; and his integrity and simplicity of manner as well as his position as archetype of the missionary-botanists mark him worthy of closer attention than this wide study can properly allow. Only a thumbnail sketch is possible.

Armand David was born as one of three sons and in comfortable circumstances at a time and place when it mattered; that is, in 1826 in Espelette, a small Pyrenean town. His father had the delightful Christian name of Fructueux, and was the local doctor, mayor and magistrate. He infected his sons with his own three enthusiasms: passing on to one a love of the practice of medicine; to another, his passion for good food, and especially game pie; and to Armand an abiding interest in natural history. In his holidays the boy would collect objects of natural history as acquisitively as a magpie, walking the Navarre foothills for eleven or twelve hours in a day, and eventually he told his delighted parents that he wished to test his vocation as a religious. Fortunately he was not obliged to choose between natural history and the priesthood. It was a simpler, less demanding age before the overspill of knowledge had forced specialization on to Western culture. Quite properly, and without any inhibitions that he was betraying either of his vocations, he was ordained and took his vows as a Lazarist in the Order of St Vincent de Paul. The Order then used his skill and made him a teacher of science.

Generally the Lazarist missionaries were sent to South America, Ethiopia, the Levant, Persia or China. David already had warm feelings for China but his superiors decided to send him to the Italian riviera, to teach science at the Savona College. Nor was it a merely temporary appointment. He was kept there for ten long years. 'All the time I dream of Chinese missions,' he wrote; but he did not resent his enforced stay at Savona. It gave him the chance to become acquainted with fellow naturalists in Europe and perfect his technique as a teacher of science. He was thorough in class, encouraging the able, catching the attention of reluctant pupils by presenting facts in an imaginative and interesting way, being patient with the tortoises and gentle and understanding with those who confessed that natural history was their blind spot. By the time his superiors did choose to send him out to China he was thirty-five years old, and he was already so well-known in Italy and France that when they heard he was going to the Far East five illustrious scientists of the Musée d'Histoire Naturelle asked the Order if he might collect for them. David was principally an ornithologist and for his work on birds alone would have won fame, but his all-roundness was of more use to science at that time. Permission was given by the Superior-General of the Lazarists and David was humble enough to count himself an exceptionally lucky man to be appointed a collector. It never occurred to him that the five illustrious scientists were exceptionally lucky to find a collector who was a first-class geologist, mineralogist, geographer, hydrographer, ethnographer, zoologist and botanist.

It was 1861. The war débris had barely been cleared before the French were building a school in Peking. The headmaster was to be a M. Morly; the staff a team of three Lazarists, David to teach science and two fellow-priests to teach mathematics and music. They travelled together from France to Peking and they became fast friends because each one had such an appetite for knowledge and could impart enthusiasm for his own pet subject. M. Morly had a love of anthropology, and David had his natural science, while the priest-mathematician delighted in physics, and the priest-musician in the manufacture of clocks. If Armand David's out-of-school collecting had not been so successful he might have remained on that happy staff for the rest of his life, but the five scientists at the Museum were so flabbergasted by the size and quality of the collections he sent home that they realized he must be properly sponsored for the honour of science and the glory of France.

The Minister of Public Instruction paid a personal call on the Superior-General of the Lazarists to ask if Père Armand David could be released from the Peking School of Science in order to lead a collecting expedition into Inner Mongolia. Again permission was granted and the surprised but gratified David set off on the first of his three noted journeys into the interior.

His preparations were sensible. He described the baggage in a diary entry for March 13th, 1866:

> It will be cold for another two months in Mongolia, so heavy winter clothes are indispensable, as well as summer clothes for the warm season, which is very hot there. It is imperative, furthermore, to carry bedding ... Add everything that is indispensable for hunting and securing objects of natural history, everything that is required for taxidermy and for herbarium specimens, boxes of all sizes, empty bottles, etc., etc. It goes without saying I have not forgotten my ecclesiastical appurtenances.

The Mongolian expedition was less important botanically than those he made to the Tibetan marches and Shensi province. In the second journey, when he was in the small, almost independent state of Mupim on the Tibetan frontier, and where the flora was richer than he could possibly have foreseen, he made the decision to collect systematically over a particular area. To the meticulously thorough collectors of later years this would have seemed an obvious thing to do but Armand David was, in fact, the first of Far Eastern plant hunters to be so methodical and painstaking. Maybe the first niggling restrictions of Mandarins had put his predecessors in the habit of collecting haphazardly wherever and whenever they could, and this they continued to do when on the edge of the world they were unnerved by the sheer immensity of China's flora. Like pigs in a rich truffle ground they rushed hither and thither, hunting plants unsystematically. David must have been strongly tempted to do the same, and as the opportunities before him on the Tibetan border were so large it speaks highly for his ideas of scientific discipline that he did not collect at random.

The same scientific approach made him a discoverer rather than an introducer of plants, and he was far more interested in the scholar's herbarium than the pleasure garden. His bag was substantial, specimens of more than 2,000 species of plants. It did not compare with the collections of other missionary-botanists who worked after him in the Tibetan Marches – Jean Marie Delavay, for instance, has been credited with personally collecting, drying and pressing more than 200,000 specimens, and the collections of Farges, Soulié, Bourdonnec, Monbeig, Dubernard, Geraldi and Boudiner were considerable as well – but David's was by far the largest herbarium to be sent to Europe in his own time and he made a point of remarking that his collection was simply a token of the Asian flora yet to be discovered. Moreover, he was simultaneously making no less than six other collections, of insects, molluscs, fishes, reptiles, birds and mammals.

In zoology he is famed for his introduction to Europe of the David Deer, *Elaphurus davidianus*, an odd creature which brays like a donkey, has the gait of a mule, and splayed feet as though its natural habitat was marshy ground. The Chinese showed their appreciation of the deer's asymmetry by calling it *Ssu-pu-hsiang*, or The-Four-Characters-Which-Do-Not-Match. It had been extinct in the wild for a century or more and in David's time only existed in the imperial park in Peking. How, precisely, he wheedled some living specimens from the park keepers is not known, but it could not have been easy though he himself made light of the matter. The deer he sent to Europe became the parents of all zoo specimens and the large herd kept at Woburn Abbey by the Duke of Bedford.

David celebrated his forty-third birthday up in Mupim, and yet he continued his boyhood habit of collecting for eleven or twelve hours at a time. And often at this time he was walking more than thirty miles a day over mountains. It was punishing to his body and he began to feel the strain. Moreover, his whole existence was lived on a knife-edge. He used to say with a chuckle that the Frenchmen's beards, quite as much as their muskets, kept off the almost hairless Chinese brigands. Anyone with less confidence in the essential goodness of mankind would not have turned his back so often on people who for generations had been taught to mistrust and even hate the 'Red Hairs' and the 'Franks'. Peking's enforced change of policy which allowed the Chinese to meet Westerners for the first time was too abrupt to be taken in and understood by provincials in the mountains. But either David was lucky, or – and this is more likely – his sense of humour and his unruffled trust were marked, for he and his property were respected in a way which even surprised the Chinese themselves. His attitude to food was characteristic, and no doubt won him friends amongst the hillpeople. He confided in his diary:

> As for food, I depend on the Chinese, and I believe that with a little goodwill, one man can live wherever another can. I do not burden myself, therefore, with carrying food, except a bottle of cognac for emergencies.

Nevertheless it was always a strain being amongst people whose ways and language were largely unknown. The terrain so dramatically beautiful, was also remarkably dangerous. Collecting in the rarified air was bad enough but there were regular and wearing dangers to be faced in the form of river travel, dacoits, avalanches, spider's web rope-bridges across chasms so deep that the noise of water torrents far below was only an indistinct murmur. Like Douglas in America, David was almost killed when his boat was smashed to pieces on a rocketing mountain river and he lost a large part of the herbarium material he had been collecting.

All these strains piled one on top of another proved to be too much for him and he fell ill. He was still unwell when he returned to the plains and, though he waited and rested for some time before making his third expedition through the central provinces to Shensi, his health broke down on the journey and he only just managed to get back to Shanghai. It was the end of collecting in China for Armand David, and in 1874 he returned to the Mother House of his Order to teach, and write up his notes, and tell traveller's tales with his accustomed modesty and charm.

He was well and affectionately remembered after his death in 1900: by his pupils as 'a handsome old man with a quick step, his face framed in longish white hair, with deep-set keen eyes under a broad and high forehead'; by gardeners for his discovery of a multitude of first-rate garden plants; by the naturalists of France for his distinguished contributions to their sciences; by his Order as one of the most famous of all Lazarists. But in his home town of Espelette, where he was still 'old Fructueux's boy', they remembered him best because on his last visit he had shown them all a spider 'as big as his fist which was trained to retrieve and answer a summons'.

5

Some Changes in the Mode of Plant Hunting

In the twenty-five years following Armand David's departure from China the manner of plant hunting on the edge of the world underwent some change. It was unnoticed by the vast majority who, in that time, had more important affairs to think about.

Fresh ideas were literally altering the face of the world. Nine more states were admitted to the Union and the United States was becoming a powerful reality. She fought and won a war against Spain and occupied the Hawaiian Islands and the Philippines. In her domestic and social life the spots and pangs of her adolescence were disappearing. True, the murder of James Garfield by a demented tuft-hunter was the second assassination of a president in office, and the romanticized cowherds in the west and the flatboat bullies who manhandled their arks and broadhorns down-river to New Orleans were unspeakably wild, but she was forging an independent culture of her own. As she shook down, the Old World was threatening to blow up. The fossilized monarchies of Europe were almost at their last gasp. And so, though they did not perhaps realize it, were the Manchus on the Dragon Throne of China. The Taiping Rebellion was eventually put down but – humiliatingly – only through the assistance of a former enemy, General Charles Gordon. He was a Scot, farouche, undoubtedly a little mad (putting up notices which forbade the peasants to drown unwanted children in his water supplies); but because he was a brilliant tactician and leader they had to honour him with the yellow jacket and the peacock's feather, and give him command of the Ever Victorious Army. Since then the Manchus had fought and lost a war against Japan and, in a further series of humiliations, had been obliged to grant privileges, rights and territories to the empires of Japan, Russia, Germany, France and England. Even before Armand David died the old Empress Dowager, whose private life was sensationally depraved and, of more moment to the history of China, whose passion for ponds and paeonies led her to misappropriate naval funds for the enlargement of her pleasure grounds, was driven out of her own capital by a Western force of Americans, Germans, Russians, French and Englishmen, and kept out for a year.

Against such dramatic events any changes in the mode of botanical exploration were barely perceptible. But they occurred.

The amateur, with one or two notable exceptions, no longer led the field. The academic botanist was becoming equally rare. In fact, amongst the plant hunters who have collected on the edge of the world only the self-taught Augustine Henry, the Swedish scholar, Dr Harry Smith, the Austrians, Camillo Schneider and Handel Mazzetti, and the American professor, Liberty Hyde Bailey of Ithaca, could be classed in this rare category. The remainder were primarily plantsmen. They collected herbarium material for botanical study but this was of less importance than the collecting of seeds and living stocks for introduction into cultivation. Plant-hunting methods also underwent a change. Haphazardly skimming the cream off the milk would no longer do. Clearly it was more practical and profitable in the end to be systematic and either make an exhaustive search for anything new over a defined area, as Armand David had done, or set out for a single objective armed with information as to its most likely whereabouts.

Fin de siècle plant hunters were more scientific than any of their predecessors and more lavishly rewarded for the pains they took. Unfortunately, though their adventures were well documented by written journals and published accounts, their portraits seldom came through. And so the introverts in their number gave an unreal appearance to the world of being strong, silent, lean, bronzed, dedicated, fearless, and with all their tendencies, regrettable or otherwise, kept strictly under control – in effect, the Hornblowers of plant hunting; while the few extroverts gave an equally unreal appearance of being sparkling and romantically eccentric. Their true characters were hidden, deliberately or otherwise, because of 'good form' before death and the principle '*de mortuis nil nisi bonum*' after it. Yet by a careful pruning of inessentials, and with regard for the disinterestedness or otherwise of those who have reported on their manner and their careers, it is just possible to see something of the real men who undertook the great plant-hunting enterprises in Central Asia after Armand David. There were many of them, and all played some part in the establishment and stocking of wildernesses, herbaceous borders and alpine gardens in the West, but a few stand out for the size or the nature of their contribution.

6

Doyen of Modern Collectors

The doyen of modern collectors on the edge of the world reached China on a June day in 1899. He had been sent from England with the specific task of tracking down and collecting David's Dove Tree, *Davidia involucrata*. In this, and on many expeditions afterwards, he was to prove so able that in the public mind his name, like that of General Gordon, was quickly associated with the country where he made such successes, and he was popularly known as 'Chinese' Wilson. He never cared for the name.

He had been born Ernest Henry Wilson in the Gloucestershire country town of Chipping Campden and the cultivating of plants interested him from an early age. His first job was as apprentice gardener's boy to a nursery firm at Stratford, whence he went on to the Birmingham Botanic Gardens. He did well enough to graduate, as it were, on to Kew, but his prospects there were not bright and he was obliged to think of getting himself trained as a teacher of botany. He did not like the idea, caring less for the schoolroom than the laboratory, less for the laboratory than fieldwork. Already he was the open-air man of the day – with clipped moustache and pipe to match. But without that slice of luck which has been vital to many great plant hunters he would certainly have gone on to teaching. As it was, Kew's Director, Sir William Thistleton-Dyer, was asked to recommend a plant hunter to the firm of Veitch and his eye fell on the twenty-three-year-old Wilson. From that day the young man's future was assured.

Veitch's instructed him for six months in the practical side of things: in the principles of plant recognition; the taking, stowing and transport of herbarium material, seeds and propagating stocks; in the use of collecting equipment and a camera. The last had already become an important article in the kit of a plant hunter and Wilson was to prove an excellent photographer, but he had the strangest fad to use always a full-plate camera. Hawking such a weighty and awkward article through the wilder parts of Asia can hardly have been easy; nor for that matter was the taking of photographs. Present-day users of the matchbox-size marvels which produce equally good results would find his insistence on a whole-plate machine, complete with tripod in brass and mahogany and large black headcloth, particularly mad.

When Wilson had finished the course, and not before, Veitch's sent him off on his first expedition to China. He was given three instructions. First he was to travel by America and not by the eastern route so that he could call at the Arnold Arboretum in Boston, Massachusetts. The Keeper, Professor Charles Sprague Sargent, had introduced many Japanese trees and shrubs to the United States and would be able to help Wilson with general advice on collecting in the Orient. Perhaps, later, Veitch's came to regret this part of their instructions because Sargent and Wilson got on so well together that acquaintance quickly ripened to a lasting friendship and Wilson ended up collecting for the Arboretum. Veitch's, though, were to have his services for two expeditions. Their second clear instruction was that on arriving in China he should make for south-west Yunnan where Dr Augustine Henry was living and ask his advice on collecting in Central Asia. This would be of particular assistance to him in carrying out their third instruction: to find and introduce *Davidia involucrata*.

Wilson enjoyed his visit to Dr Henry in Yunnan. He was there two months, botanizing with his host and learning a great deal from him about local conditions. Then, at Henry's suggestion, he set off for the area round the gorges at Ichang on the Yangtze in search of the elusive Dove Tree.

That the first quarry of the first modern collector should be the Davidia or Dove Tree showed how systematic plant hunting had become. Armand David had first found the tree up in Mupim and had described it enthusiastically both because it was beautiful and because he guessed rightly that it stood on its own in the vegetable kingdom, its nearest ally being the Nyssa, and was, therefore, a scientific curiosity. He had made up herbarium specimens of some of its parts, but being a scientific collector this was sufficient for him and he did not even try to introduce the tree to France. Paul

Farges, another missionary-botanist who was in China after David, also found the tree, and in 1897 he sent a packet of thirty-seven seeds to M. Maurice L. de Vilmorin who had made a huge botanical collection of Chinese trees and plants in his pleasure grounds at Les Barres. Whether the directors of Veitch's and Ernest Wilson knew of this is uncertain. They could not have known that out of Père Farge's thirty-seven seeds only one germinated and that it actually put out its shoot on the June day when Wilson landed at Hong Kong.

The Davidia was a tree worth looking for. Since that time it has been called a number of names – the Dove Tree, Pocket Handkerchief Tree, and Ghost Tree – all on account of the pairs of white hood-shaped bracts of uneven size which in May are shown off to perfection by the tree's aromatic leaves of a vivid green. They could be said to resemble doves gone to perch or, at a pinch, pocket handkerchiefs on a linen line; and as they flutter down from the branches the bracts do give a ghostly effect.

Wilson scoured the countryside around Ichang, moving slowly west and into the Land of the Four Rivers which was to become his special collecting territory, and eventually he found what he had been sent to collect. Wilson's luck was proverbial. On this occasion he not only found a colony of Davidias but besides them some Tetracentrons, a genus which had been discovered and described by Augustine Henry but had yet to be introduced to the West. *T. sinensis* was added to the expedition's bag. The Davidia, though, commanded all his interest for the moment.

He had been brought up to be thorough and trained in exactness by a Veitch foreman. The first essential at that exciting moment was to establish that the small colony of small trees in front of him really were Davidias. In all particulars save two they corresponded with Armand David's notes: but whereas he had described the underside of the leaves as whitish and felted, the underside of the leaves before Wilson were a yellow-green frog colour and perfectly smooth. Clearly, though, they were Davidias whatever the species.

In a diary entry dated May 31 1900, Wilson wrote:

. . . Ascending a precipice with difficulty, we soon reached the Davidia trees. There are over a score of them growing on a steep rocky declivity; they vary from 35 to 60 feet in height, and the largest is 6 feet in girth. Being in a dense wood they are bare of branches for half their height, but their presence is readily detected by the numerous white bracts which have fallen and lie strewn on the ground.

Later he would come back for seed, a hard, winged nut contained in a pear-shaped fruit the colour of a damson. Meanwhile he proposed to make a permanent record of the discovery. We are reminded with a jolt that he actually had that massive full-plate camera with him in the wilderness. He decided to photograph the Davidias in their full May glory. On a steep slope in a dense wood, and with a camera as unwieldy as an orange box on legs, the task was formidable. But Wilson was a practical man and a great improviser. He noted that one of the Tetracentrons was growing on the edge of a cliff overlooking the Davidias. Two of the party joined him in scaling the tree and chopping away branches to make a clear space so that he could photograph the nearest Davidia. Pulling up a woodsman's axe on ropes, and afterwards that unmanageable camera, was difficult enough. Perching behind the camera with head shrouded in a black cloth while focusing on an upside-down Davidia was even worse. Worst of all, as Wilson found in climbing the tree, the wood of a Tetracentron is exceedingly brittle and he commented in his diary:

> ... this does not add to one's peace of mind when sitting astride a branch about 4 inches thick with a sheer drop of a couple of hundred feet beneath.

It was disconcerting, too, not to be absolutely sure whether or not he had found David's tree, and years would pass before he could know. In fact M. Vilmorin's single seedling grew to flower as a tree at Les Barres in 1906, and this differed from the others in that the underside of its leaves were hairless but a glaucous colour. It was listed as *Davidia vilmoriana*. The seed Wilson sent home in quantity from his first expedition, that is from the trees with frog-green, hairless leaves, grew into trees which were named *D. laeta*, and on a later journey in China he took seeds of Armand David's original discovery, *D. involucrata*.

Typically, Wilson had eyes and interest for nothing save his quarry. At the very time he was taking photographs from a high and insecure perch in Szechuan, the anti-missionary and certainly anti-foreigner Boxer tong was murdering and looting in Shantung and other northern provinces. The demands of the Treaty Powers had become excessive; their open contempt for all things Chinese particularly offensive. In the Shanghai park for instance, which was largely ruled by British interests an insulting notice declared 'Dogs and Chinese not allowed'. Inevitably there was a reaction from those Chinese who found the constant humiliations unbearable. The Empress Dowager and the Mandarins at first mildly opposed the Boxers but then openly supported them. The German Minister at Peking was knifed, several of the foreign legations were ransacked and burnt, and upwards of 200 foreigners took refuge in the British legation where they withstood a siege until a Western expedition arrived by ship, blew the Chinese forts to bits, and advanced steadily to their relief.

It was less than safe to be a 'Red Hair' or a 'Frank' anywhere in China at that time, but there was Ernest Wilson, obsessed with plants, apparently indifferent to menacing tongs, and vexed because the political situation obliged him to keep within a certain distance of Ichang.

Having found a Davidia, if not the Davidia, he filled in the time while waiting to return for seed by searching for other plants. He also got himself shipwrecked on the Yangtse-Kiang when his river sampan – mixture of barge and houseboat – was snagged on a boulder and turned turtle. A good many of his belongings, including the full-plate camera and hundreds of exposed and unexposed plates, went to the bottom. Wilson cursed, got himself ashore, replaced the camera as soon as he could, and regarded the disaster as a good puritan should; that is, as a sort of divine chastisement to balance against his undoubted good fortune while hunting plants. Though English, he had many of the characteristics of Scottish collectors, being puritan and suspicious of pleasure and prizing hard work and being provident as spiritual virtues. He shared, as well, their dourness and their thoroughness. The methodical fashion in which he hunted plants almost always brought him success; and his thoroughness applied to an area where a large variety of temperate plants grew in great abundance was bound to have a startling effect.

From his first expedition alone when hemmed in by Boxers he produced not only what he and everyone else thought was David's Dove Tree but also the seed of 305 other plant species, 35 Wardian cases of tubers, corms, bulbs, rhizomes and rootstocks and the dried and pressed herbarium material of 906 plant species.

Needless to say he had hardly been in England for six months before Veitch's sent him out again: on this occasion to aim principally at the Yellow Chinese Poppy or Lampshade Poppy, a Meconopsis (*M. integrifolia*) with Claude-tint yellow, cup-shaped flowers a hand-span in diameter and topping three feet in height. Antwerp Pratt had discovered the plant and promised that it was one of the most striking in all the Chinese-Tibetan frontier lands. For this trip he based himself at Kaiting-fu, which lies much farther west than Ichang and is in the heart of Szechuan. Near-by is precipitous Mount Omei, one of the three holiest Buddhist mountains in China and peppered with shrines and temples, and this with two other mountains makes a triangle on the map of an area which is so rugged that the Chinese call it the Laolin, or wilderness. It is limestone country, pocked with crags and gulleys, and is drenched for the greater part of the year, and Wilson discovered it was an Aladdin's cave to a plant hunter. There he found Antwerp Pratt's Lampshade Poppy, and another Meconopsis, *M. punicea*, smaller and scarlet but no less a beauty; and also the famous Rhubarb, *Rheum alexandrae*; several unknown Primulas and Rhododendrons (despite the fact that their feet were in limestone); two favourite species of Roses, *R. moyesii* and *R. wilmottiae*, and a Poplar, *P. lasiocarpa* which though it is not often seen today has one of the most handsome of all foliages. For the Poplar alone all Wilson's efforts would have been worthwhile. His puritan sense was gratified by the fact that he had, so to speak, to 'pay' for his successes with fatigue, discomfort, the perpetual threat of being deserted by his coolies and, on a never-to-be-forgotten occasion, even with starvation. He was far away from Kaiting-fu and took an unknown and almost disused track through difficult country in appalling weather. The party had insufficient food and the few people they met were themselves dying with hunger because the harvest had failed. Wilson and his men only just escaped with their lives.

After two expeditions on behalf of Veitch, Wilson bound himself to Professor Sargent and the Arnold Arboretum. It was a sensible move. The firm of Veitch was running down. Its financial resources were evidently limited. Its directors were coming to the ostrich conclusion that almost everything worthwhile had already been collected in China. As a result their instructions to Wilson were precise and his activities hedged about with financial restrictions. Working on behalf of an American university gave him larger scope and a greater freedom than he had known previously for, although he was naturally expected to hunt out ligneous plants, he could do so wherever his knowledge and instinct told him to go.

The outstanding success 'paid for' with hardships was a pattern that was to repeat itself throughout most of his collecting career. Except on his last and most leisurely trip to the Far East when he introduced the Taiwania from Formosa and saw, for the first time, Kurume Azaleas in Japan, Wilson regularly met with taxing adventures. His very near escape from death under an avalanche when he gained his 'Lily limp' has already been described on page three. There were other episodes as well which would have worn away a less robust man. In his published writings Wilson was either too technical and remote or too chatty to show himself very clearly, but it seems that he met, accepted, and then shrugged off the difficulties and dangers as part of the job. In compensation no one has ever collected so many fine plants, both woody and herbaceous,* and when his plant-hunting days were finished he continued to enjoy good fortune and attract disaster right to the end. He was made Assistant Director at the Arnold Arboretum and then his friend Sargent's successor as Director. As if to match or 'pay for' the enjoyment of so superlatively suitable a post, he and his wife were killed in a motoring accident in 1930, on a highway near Worcester in Massachusetts.

* Professor Sargent proved the extent of Wilson's accomplishments by writing an account of his introduction of woody plants to the Arnold Arboretum. It was entitled *Plantae Wilsonianae*, and published in nine parts forming three large volumes.

An Economic Botanist

Institutional collectors was the name given by E. H. M. Cox to those who hunted plants for a botanic garden or government. Clearly they were mainly restricted to the particular interest of the institution they represented but, as 'Chinese' Wilson found when he changed from a firm of professional nurserymen to work for the Arnold Arboretum, they enjoyed a certain amount of liberty. Indeed they were expected to be enterprising. Wilson's contemporary, the American Frank Meyer who worked for the United States Department of Agriculture, was one of the most enterprising and original institutional collectors of his day.

By birth Meyer was a Dutchman and had been an assistant to Hugo de Vries. But he was no laboratory botanist and a hobo at heart. He not only adored travelling but he preferred to do so on foot, and even at a time when walking was a possible and pleasurable exercise and when by inclination there were many hikers in the United States, Meyer's energy and gusto as a walking explorer-botanist became well known. He was actually striding through the southern states and Mexico when the Department heads selected him to be their economic plant hunter in Eastern Asia.

The post was entirely new, created to meet a demand at home for greater varieties of grain, fruits and vegetables. These were surprisingly limited considering that North America contained four floristic regions and eight climatic provinces. Reports from the opened-up interior of China suggested that her agriculture, though primitive, was robust. And in far-away Russian and Chinese Turkestan there might well be many hardy strains of seed best suited to the colder parts of the United States. As for the rest, the Department's collector was required to use his initiative or, as they put it then, his head.

No appointment could have pleased Meyer more. Like Joseph Hooker he enjoyed romance as well as the practice of science, and, being so passionately devoted to detail that he would spend hours investigating and describing the habitat of a single plant, his awareness of its other attributes could easily move him to hail it in an ode. In Eastern Asia he could indulge himself in both romance and science, and walk and walk and walk. Furthermore, he would be in one of the largest and most dramatic regions on earth, and paid for it into the bargain.

Meyer made four expeditions for the Department, going as far west as Russian Turkestan and travelling extensively in the east: right through the great Chinese plains, Manchuria and Korea, up by Lanchow on the Great Wall and into Kansu. Unlike Caesar he did not keep to winter quarters but spent the time inspecting grain and, as their sap was down, taking fruit tree scions. He found new varieties of rice and beans and fodder plants as well as fruit and vegetables, and his total work for the Department was enormously important to husbandry in the United States. Farmers from Cape Flattery to Florida and from Fundy Bay to San Diego ought to keep an annual Meyer feast in grateful remembrance of his work in Asia. Gardeners, on the other hand, have scarcely heard of him. In the main his introductions were economic plants, concerned far more with the pot than with pleasure grounds, though this is not to say economic plants have no aesthetic appeal. On the contrary. Some of Meyer's soya and other beans, and certainly his fruit trees had a distinctive attraction and beauty of their own. And as Meyer tramped from place to place and from clue to clue in trying to trace a particular strain to the centre of its distribution, though he always had one and a half eyes open for other economic plants, half an eye was spared for solely decorative species.

He re-found and introduced the single yellow Rose, Canarybird (*R. xanthina*, a discovery of Armand David's more than thirty years before); *Actinidia kolomikta*, a fragrant, white-flowered climber, and varieties of wild Plums, Peaches and Almonds, Viburnum, Juniper and Lilac. Three of the shrubs he introduced, the Chinese Spindle Tree, *Euonymus bungeanus*, *Syringa villosa*

and *Kolkwitzia amabilis*,* only did well in the States in areas where
the seasons are well defined. In Eng'and sharp and unseasonal
frosts cut them back and they never could reach their full size or
show off their full beauty. The list is by no means complete and for
a plant collector who could spare but half an eye for ornamental
plants and who often travelled during wintertime when identifying
sleeping trees and shrubs was difficult, it was something of an
achievement. No doubt, as well, it would have been more impressive still if his collecting career had not been cut short.

His last expedition took him over the frontiers of Wilson's
country. After wintering near Peking, making forays deep into the
surrounding countryside, he went south across the Great Plain of
China to the old Treaty Port of Hankow on the Yangtse-Kiang. It
was an important trade centre and ideal as the headquarters of an
economic plant hunter because amongst its main exports were tea,
silk, a grass-cloth made from *Boehmeria nivea* fibres, sesame seeds,
beans, tobacco, timber, vegetable oils and Chinese herbal medicines.
In the vegetable market alone there would be a large number of
species differing in characteristics from those grown in the United
States. Chinese Peppers, for instance – that is, the pulverised fruits
of *Zantholxylum bungei* – were grown on everyone's cabbage-patch
but were only grown for export in the Min Valley far away upstream from Hankow. And from the same area, especially the plain
of Chengtu, came Capsicums and Chillies and salted shoots of the
Bamboo *Sinarundinaria nitida*. From the warm south came supplies of dried Bamboos. From overseas, for distribution to other
markets in the provinces, came junkloads of an unusual vegetable,
the seaweed *Porphyra vulgaris*. There would be mounds of roots
and flowerheads and stem vegetables and seeds and leaves, dried
and fresh and salted, all mostly unfamiliar but not unpleasing to
Meyer's Western palate.

Once he was on the scent of an economic plant which might be of

* The Syringa and the Euonymus had in fact been introduced to Kew
in a consignment of seeds sent there by Dr Bretschneider where they
grew to be half-pint specimens; and the Kolkwitzia was introduced to
England by Wilson in 1901 though it never became popular. With many
of his other introductions cultivated at Veitch's Coombe Wood nurseries,
the Kolkwitzia came under the auctioneer's hammer just as it was about to
flower for the first time and no effort was made to propagate it. Meyer's
introduction of all three plants was on a large scale and permanently
effective.

value in the States no bloodhound could have been more determined in following it. On this occasion his nose took him upstream
from Hankow to Ichang. He established a base and made a thorough
survey in surrounding Hupeh for the larger part of twelve months.
Then something happened, but to this day the facts are hidden.

At the beginning of June in 1918 his swollen corpse was found
floating in the sluggish waters of the Yangtze below the port of
An-king. Perhaps he had been on his way downstream making for
Shanghai and then America. But no one could be certain.

The Yangtse was sometimes the plant collectors' friend. She
had carried professionals, missionary-botanists, consular officials,
and 'honoured commissioners of Customs' up into the high western
provinces. Upon her, Antwerp Pratt had sailed in state with his
German assistant, Kricheldorf. But she was not always gentle;
claiming Wilson's equipment and almost Wilson. Now she had
swallowed up Frank Meyer.

8

The Professor of Chinese and Botany

Vienna – for 500 years a centre of botanical interest – was the birthplace of Joseph J. Rock, who may be accounted the most distinguished of America's institutional collectors. He was something of a prodigy, being well schooled in science and many languages at a stage when most boys still have their interest focused on the football field, and in every way he was suited for the rather quaint chair of Chinese and Botany to which he was appointed at the University of Hawaii in 1907. In 1913 he became an American citizen and in 1920 economic plant collector of the Department of Agriculture.

Rock's first mission was to track down *Taraktogenos kurzii*, a tree called the Kalaw by native tribesmen and of large medical importance. The adventure was invested with fantasy right from its beginning.

Imaginary cures for leprosy often featured in Eastern folklore, and the administrators of British India were understandably cynical when miracles were accredited to a tree which grew high in the wet forestlands of Central Asia. The account was too whimsical altogether. It even began like a fairy story: 'Once upon a time there lived a king of Benares called Rana, and Rana was a leper . . .' The tale followed a romantic and fantastic theme. Rana, bound by his own laws, honourably cast himself out and, not wanting to be seen, he hid himself in the wet forestlands. There he lived in a hollowed-out tree called the Kalaw and he ate the tree's fruit. Soon he found that the hated white patches on his skin were disappearing. Within a month he was entirely cured. To add a relish, the story continued that in a near-by cave lived a princess who also had leprosy, and that Rana cured her with Kalaw fruits, married her, gave her sixteen pairs of twin sons, and not one of the thirty-two had leprosy. Everyone lived happily ever after.

It seemed to be against reason to suppose the story had any basis in fact. But in Indian markets a foul-smelling, dark-coloured oil called *chaulmoogra* was sold at dizzy prices because it was claimed to cure leprosy. And this oil, said the vendors, had been brought down from the high forestlands of Central Asia where it was pressed from Kalaw fruits. Though it appeared mad, medical officers could not afford to ignore any means of trying to cure leprosy and they made tests. To their astonishment they found a drug in the oil which, though it hardly worked miracles, did stay the disease and was particularly useful in its treatment. The administration and world medicine were impressed. The discovery made it imperative to start Kalaw plantations without delay because, clearly, eight or ten years would have to pass before the trees were sufficiently mature to produce oil.

The largest snag in proceeding with the plan was that no one in the plains had the least idea what a Kalaw looked like. Bazaar merchants produced seeds which they said were of the Kalaws but ten years later, when the seeds had germinated, grown into trees, matured and given their first lot of oil, they were found to be no such thing. If they were Kalaws they belonged to a different species from the Kalaw which produced chaulmoogra oil. The true Kalaw remained undetected until 1890 when a botanist sent in seeds to the Director of a Botanical Survey in India and they were identified as the bearers of genuine chaulmoogra. But the absurd frustrations continued. The seeds had been sent from Chittagong but no one had troubled to note their place of origin.

On the basis of this single clue, various attempts were made to trace the Kalaw, but they met with no success. And yet expensive chaulmoogra oil still found its way through the jungles to the bazaars of Calcutta and from there was distributed to western drug dealers.

In 1920 officials of the U.S. Department of Agriculture reassessed the chances of tracking down the tree. It was in that year, too, that Professor Rock resigned his chair of Chinese and Botany and took his exceptional gifts into their service. He was instructed to find the Kalaw and succeed where for so long others had failed.

He aimed first for Siam, determining to move north from there as far as the trail might lead him. His only consolation was the undoubted fact that the tree existed and did produce chaulmoogra; otherwise, so fantastic was the hunt, he would have been tempted to give up. In Bangkok all his inquiries met with the same reply. Yes, everyone knew about the Kalaw. Everyone. But, no one knew where it could be located. No one. A clue – merely the girlhood memory of a viceroy's wife – took him to Korat. There he was told that his quarry would be somewhere deep in the jungle. Fruiting trees had to be a long way from trails or villages because the saplings made fine firewood and Kalaws near villages never survived to mature. Further clues took him on a 350-mile trek across Siam to the hill city of Chiengmai. It was a false trail, but the next, by houseboat downriver to Raheng, was much more promising. When this turned out to be another disappointment Rock almost turned back. Anyone less dogged would have done, but a short rest and an interesting piece of bazaar gossip put new life into him. He followed the trail right over the mountains between Siam and Burma, having his first taste of leopards, tigers, and a local and particularly virulent brand of snake. But in compensation, he at last found a species of Kalaw. A few trees. But either they were mules or the season had been a bad one, for none of them carried any fruit. Rock went on to Moulmein in the Gulf of Mataban and then up north into the hills. He came across one single fruiting tree. But this was not the *Taraktogenos kurzii* he had been commissioned to find but a closely related species, and though he managed to collect 170 seeds he found himself in strong competition with monkeys who ate the fruits and threw the seeds away and with porcupines below who went for the seeds as crazily as a hound will follow aniseed. He pressed on northwards, following clues, rumours and speculations until near a filthy village on the upper Chindwin River he found a colony of the true Kalaws. It left a good deal to be desired. Many of the trees had been lopped for kindling and the villagers had ringbarked others to make an infusion with flakes of the strippings. He could only collect a very few seeds. But he was obviously on the tree's trail. In the very next village he found more Kalaws, and finally his small party was 'loaded with seeds and could carry no more'. At this point a man-eating tiger began visiting the village where they had their headquarters to seize a villager a day. Then a storm and a rogue elephant put paid to the village altogether. Rock and his men were lucky, escaping the depredations of tiger, tempest, and elephant, and they had sufficient seeds for plantations of Kalaws to be established in American possessions in the Pacific and so assure constant supplies of pure chaulmoogra. It was a dramatic end to one of the most arduous of all plant hunts when the quarry seemed to belong more to the realm of myth and magic than to scientific fact.

The pattern of Rock's collecting career thereafter was much wider than that of other modern plant hunters. His principal collections were of plants, but he also hunted birds and animals, and in addition to working for the U.S. Department of Agriculture he collected for the Arnold Arboretum and the Harvard Museum of Comparative Zoology, and he twice made expeditions under the auspices of the National Geographical Society of Washington. The seeds and plants he collected on the last expeditions were syndicated to private gardeners and commercial nurserymen both on the Pacific coast of North America and in Great Britain.

He travelled widely: in Yunnan, where his collecting crossed that of George Forrest and Frank Kingdon-Ward; in the small and partly-independent state of Muli, where he found few plants but made great friends with the king; and in the land of the Ngoloks, a belligerent tribe of Tibetans who lived by the Amne Machin. This range of high peaks, separated from the Min Shan by the most turbulent and untamable of all rivers, the Hwang Ho, was on the roof of the world, and with its nine-month winter and three-month summer it had little of value for Rock in the way of plants: but he called it 'one great zoological garden ... wild animals grazing contentedly ... deer, wapiti, and many others'. He enjoyed himself most, though, in what he called Tebbu land; that is, the country peopled by the Tepos south of the Min Shan which is divided into two by the Satani Alps.

Between Vienna and the Land of the Ngoloks, Rock managed to see more of the world than most men. It was Tebbu land which fired his imagination:

> I have never in all my life seen such magnificent scenery. If the writer of Genesis had seen the Tebbu Country he would have made it the birthplace of Adam and Eve.

Finally Rock made an expedition which in some ways was as fantastic as his first in search of the Kalaw. This was in a wild area to the west of Muli in Szechuan which had once been under the control of the Tibetan Prince of Litang. Since the Chinese-

Tibetan frontier quarrels the region had become wilder and wilder and by 1928, when Rock wanted to collect there, it was entirely lawless and in the hands of a rogue called Drashetsongpen who had turned from the holy life of a lama to the excitement of being a bold robber chieftain. Fortunately for Rock this larger-than-life bandit was a friend of his old friend the King of Muli, and the American had special permission to botanize throughout his wild domain. He enjoyed his visits and planned to go back there to collect seed, but when bad conditions ruined the grain and Poppy harvest the peasants were on the lookout for someone to blame. A white stranger wandering in the hills, picking flowers and irritating the gods would do. A polite message was sent from the robber's stronghold to the royal court at Muli to suggest that, in view of the circumstances, it would be imprudent for Mr Rock to return. And Mr Rock did not.

The Prince of Alpine Gardeners

Reginald Farrer and Henry John Elwes were amateur modern
plant hunters amongst a large number of professionals. And
both had colourful personalities. No more than Banks in
the seventeenth century could either of them settle to enjoy the
privileges of their birth as Edwardian country gentlemen. Elwes
was from Gloucestershire, Farrer from Yorkshire. Their common
taste for collecting was motivated by very different reasons.

Elwes retired from a Guards regiment at the age of twenty-four
because he considered peacetime soldiering too cosy and moribund
for a man of spirit. To slake his thirst for adventure he took to field
studies of zoology, ornithology and botany. Being very rich he was
able to travel extensively and he would, literally, leave at a moment's
notice if he had news of a rare plant to collect or a rare bird or beast
to shoot and stuff. On their trail he collected in China and Japan,
Siberia and Russia, Tibet, India, Turkey and the Americas. He was
not, though, simply a slaughterer and bagger of rarities. He became
learned in the sciences which interested him, especially botany.
His collaboration with Augustine Henry in the seven-volume
work *The Trees of Great Britain and Ireland* and his illustrated
Monograph of the Genus Lilium were evidence of his scholarship.

Reginald Farrer was equally learned in his own department, and
a plant hunter of larger importance. He was not especially interested in adventures; nor, being short and portly, was he the sort of
man to give vent to his high spirits by taking violent and dangerous
exercise. On the contrary, his interest in botany began because as
a boy he was delicate. He had to be tutored at home and an inevitable part of such an informal education would be nature rambles in
the near-by limestone hills and dales. Farrer's imagination was

stirred by the thought of finding a plant there which no one had
ever listed before. In the 1890s, even in Yorkshire, this was aiming
high. But he was successful, discovering a new species to Great
Britain, *Arenaria gothica*.

At the age of fourteen he had a rock garden of his own and had
begun to serve that apprenticeship which one day was to make him
the prince of all Alpine gardeners. Discovering that an impoverished mixture of stone chippings plus a modicum of loam or leaf-
mould were best suited to Alpines, he found that he could grow
plants from high altitudes with much greater success than those
who indulged them with composts and stimulating chemicals. His
method of constructing rock gardens also guaranteed that the plants
had the light, air, food and water which they needed.

To begin with, then, he was a practical gardener and a collector;
but after travelling widely – in Japan, where like many before and
since he lost his heart to the Japanese and their way of life,* and in
the European Alps which he turned inside out in his search
for plants – he became well known as a writer. Had he never made
any plant introductions at all his name as a writer on plants would
have been assured a place amongst gardening *cognoscenti*, and his
views on Alpine gardening very soon commanded high respect. He
fulminated against the average Edwardian rockery, and he poked
fun at the prime showpieces of the day which had been constructed
at Henley by a solicitor whose recreations were listed as 'company
law, horticulture and microscopy'. This monster rockery was
studded with caves and had an imitation Matterhorn done in slate
and concrete complete with herds of tin chamois. Farrer's influence
in altering the style of rockeries was considerable and his classic,
The English Rock Garden, remains a perfect bedside book and is
much sought after.

During his preliminary researches for the book he came across
references to Przewalski's collecting in the north of Kansu and to
Potanin's in the south, and because the province seemed to be
typically Alpine it was a natural magnet to him. He set off for
China and arrived in 1914.

* Not, though, to their cuisine. He had a word of praise for a farthing
bun – 'light and misty in composition and her flavour is of toffee, strangely
subtilised' – but he disliked raw fish with gills pullulating on the plate to
prove the freshness, and he particularly resented the loss of a favourite
kitten which was served up to him at dinner in a species of *béchamel*.

Farrer's expedition in Kansu where wheelbarrows were the main form of transport was made with William Purdom, a professional collector who had worked for Veitch and the Arnold Arboretum. He enjoyed it immensely. As an alpinist he found the flora absorbing. As an imaginative writer he was delighted to find that one of their camps happened to be in the territory of a mad bandit called the White Wolf. As a plantsman he was much interested in the diversity of crops caused by the variations in climate throughout Kansu. He saw Wheat growing, and Dates, Beans, and Tobacco, Grapes and Onions, and Rhubarb in great quantities. Finally, as a man of keen sensibility he found the countryside spellbinding. The province contained hundreds and hundreds of miles of the Great Wall, and like Gaul was divided into three parts: the southern river valley, the eastern plateau of loess criss-crossed by ravines, and the north-western section hemmed in between mountains and desert. The separating ranges were awe-inspiring. He described the mountains as 'stretching across the world from easterly to westerly in one unbroken rank of impregnable eighteen-thousand-foot dolomite needles, crags, castles, and pinnacles'.

The outbreak of the First World War in August completely upset the expedition. Though he knew he was medically unfit for active service Farrer wanted to hurry home and offer to help in any way he could. The expedition was cut short and a rush made back to Europe. As a result many of his plants were lost in transit and the few which did get through had to endure the neglect imposed on gardening affairs in time of war. Not many survived. Amongst those which did were the nippy Threepenny-Bit Rose, *R. farreri*, the sweet-scented Edelweiss, *Leontopodium haplophylloides*, the Silver Geranium, *G. farreri*, and *Buddleia alternifolia* which Farrer described in its native habitat as sweeping 'in long streaming cascades from all the loess cliffs like a gracious, small-leaved weeping-willow when it is not in flower, and a sheer waterfall of soft purple when it is'.

Farrer's war was spent by day at the Ministry of Information, and by night writing. He added to and corrected *The English Rock Garden* and wrote two books about Kansu, *On the Eaves of the World* and *Rainbow Bridge*. And as soon as peace came he was off again to the Far East to carry on collecting. This time he had as his companion E. H. M. Cox, who clearly respected his skill as a plantsman and was gentle with his whims.

Their destination was the result of elimination. Farrer did not want to return to Kansu. He longed, instead, to go to Nepal, but the political situation ruled that out. Szechuan and West Hupeh were exclusively Wilson's country and he wasn't generously disposed towards poachers. Likewise, Yunnan was being methodically picked over by a Scotsman, George Forrest. The only region which offered a whole scalp in 1918-19 was Upper Burma, and there Farrer went with Cox.

The expedition brought in pelf, no more. After a preliminary trip Cox had to return to England, taking with him a vivid recollection of Farrer in the field:

> ... his stocky figure clad in khaki shorts and shirt, tieless and collarless, a faded topee on his head, old boots, and stockings that gradually slipped down and clung about his ankles as the day wore on. The bustle of the early start; the constant use of the field glasses which always hung round his neck ... his intense satisfaction when a plant was once in the collecting tin and was found worthy; his grunt of disapproval when it was worthless; the luncheon interval with its cold goat rissole and slab chocolate; his enjoyment of our evening tot of rum, a necessity in the rains; and, above all, his indomitable energy that never spared a frame which was hardly built for long days of searching and climbing.*

Farrer was altogether an improbable character. There has never been an alpine gardener like him; nor a plant hunter so dashing and individualistic. Though indifferent, as Cox wrote, to appearances out in the field, in London society he was something of a dandy and not a little vain of the drooping black moustaches which gave him the look of a Mexican generalissimo. His manner of pronouncing proper nouns in his own way anticipated Churchill's idiosyncrasy and as Chinese is not easy to catch precisely, he made it much more difficult for himself and everyone else by marking his maps and writing names in his books as they sounded phonetically to him. With all this, he was a kindly and cultivated man, a reader of metaphysical poetry and never without a complete set of Jane Austen even in the trackless regions where wild flowers grow.

Though Cox had had to leave him, Farrer decided to continue collecting in High Burma, and set out on his plump pony (improb-

* E. H. M. Cox, *The Plant Introductions of Reginald Farrar* (London, 1930).

ably called Spotted Fat) accompanied by a faithful servant (improbably called the Dragon). His days were much the same as before; hunting out plants and taking seed whenever it was possible and he considered it worthwhile. He had a habit of disregarding plants which, within his expert opinion, would not tolerate transportation to an alien soil or were not worth the trouble. As a result his herbarium, though put together with meticulous care, was a plantsman's not a botanist's. The dried plants were records of species from which later he intended to take seeds or propagative parts. Trying to dry them in the sodden Burmese jungles was an unenviable, next to impossible task.

Farrer did his best. He invariably did. But although he was only forty the climate caught at his chest and wrung out his strength. He could do less and less each day, and faded before his coolies' eyes. Unable to eat anything, he was reduced to a diet of whisky and soda. At length, 8,000 miles from home, in the unrelenting rain of the Chinese-Burmese borderlands he died; 'without' – as the Dragon wrote to Mr Cox – 'giving any pain or trouble to us'

10

South of the Cloud

Two twentieth-century professional collectors deserved a Roman triumph from gardeners all over the world for the number, quality and character of their introductions from Central Asia. One, Frank Kingdon-Ward, typified the widely-roving explorer who collected wherever he happened to be travelling or who on occasions set out to search for a particular plant. The second, George Forrest, typified the systematic plant hunter who combed district after district with such thoroughness that it soon became known as his.

Frank Kingdon-Ward shared with Joseph Hooker the inestimable benefit of being born into botanical circles. His father was Professor of Botany at Cambridge and many friends of the family were academic and field botanists who stimulated the boy's inquisitiveness by talking shop. Very quickly he showed an insatiable appetite for exploring, and would go to great lengths to gratify it. Simply to get to the Far East he took a schoolmaster's job out in Shanghai, and there broke his three-year contract because he had a chance of going up into West China with the American zoologist Malcolm P. Anderson. He afterwards described this memorable journey by junk and on foot as an adventure across the breadth of China 'from the coast to the edge of the world', and it inspired him to set up as a professional explorer-writer-photographer-botanical collector. Like Frank Meyer thereafter he led a full and happy life doing precisely what he wanted and being paid for it into the bargain. His journeys provided him with material for travel books and articles, photographs, herbarium material for botanical institutions,* and – most profitable of all – seeds and

* His dried specimens are now in Cambridge, Edinburgh, London, Chicago, New York and Gothenburg.

plants for gardeners. At first working for a nursery firm, and afterwards for private patrons or syndicates of gardeners, and once for the New York Botanic Garden, he hunted plants in parts of Yunnan, Szechuan, Upper Burma, French Indo-China and Southeast Tibet, as well as gleaning after Hooker in the Assam Himalaya, and between 1909 and 1957 he introduced hundreds of new species into cultivation. His largest bag was in Rhododendrons and Primulas, but he also hunted down trees for coffin timber ('to the Chinese the best thing to look forward to in life is a first-class funeral'), some remarkable Lilies and Gentians, Pitcher Plants, Tea plants and the famous Blue Poppy of Tibet. He added new names and contours to unmapped regions and taxonomists have testified to their high regard for his work by using his own name for two genera, *Kingdon-Wardia* and *Wardaster*, and for no less than forty-two specific names. It is not surprising that with Shackleton and Scott, Teddy and Kermit Roosevelt, Whymper, Sven Hedin and General Pereira, Kingdon-Ward was very much a schoolboy's hero of exploration before the Second World War. At the age of seventy-two he was still working and actually planning two more plant-hunting expeditions, one to North Persia and the other to Vietnam, when suddenly and unexpectedly he died. It seemed unbelievable, for he had that weathered, sinewy quality which gives a man the appearance of being immortal.

Confident, competent and bold, Kingdon-Ward had a distinguished and rewarding career. Because no man accustomed to command could show for ever the self-effacement of an Oriental, he was sometimes charged with vanity, and because success bred envy he must have had his enemies. But any strains and stresses he experienced he decently kept under, which made him appear a very uncomplicated, friendly man. The botanical explorer's cap he chose to wear fitted him very comfortably.

George Forrest, the methodical comber of districts, was different: a solitary made brusque by his lack of confidence and as self-punishing as all stubbornly independent people. In the course of twenty-eight years he sent home from the high province of Yunnan no less than 31,000 herbarium sheets and an equal number of separate parcels of seeds: a demanding undertaking in which the physical and mental dangers and discomforts, with the intolerable tediousness of so much that had to be done, showed what the trained human frame and spirit could withstand. Forrest's were well trained. His personality had been a good deal forged by the bleakness of his background, by the loneliness and the unhappinesses of his youth, and by the hard business of maturing on a shoestring. For this reason his character seems to be more interesting than Kingdon-Ward's and his career is chosen to conclude this short account of modern plant hunting.

With David Douglas, Robert Fortune and many of the greatest collectors Forrest was a Scot of humble birth. He came from Falkirk in Stirlingshire though he went to school on the other side of the Lowlands, in the Ayrshire town of Kilmarnock. His first job in articles to a pharmaceutical chemist was not to his liking, though his meagre knowledge of medicine and drugs and his willingness to practise a primitive form of surgery was very useful to him later in Central Asia. As part of his training he had to study botany and collect local plants. But this, too, was not to his liking. He was another reluctant botanist with no more inclination to hunt plants than Augustine Henry. His twofold aim was to better himself and, if he could, work out of doors. Pharmacy could not hold him. He went off as a lad to Australia to work as a rouseabout in the bush, and he remained there several years. The work was out of doors but within the narrow standards of his class and background the life of fossickers and rouseabouts down-under was scarcely respectable. He, with characteristic phlegm, said nothing about it at all. It is doubtful if he collected plants or took any interest in botany out in Australia but when he returned to Scotland in 1902 his half-apprenticeship in pharmacy was useful to him. He applied for work in the Edinburgh Botanic Garden. Professor Sir Isaac Bayley Balfour interviewed the stocky little man whose weather-beaten, unsmiling face was rather daunting, and he apologized. He was very sorry, but the only available job was a humble, clerkly one in the herbarium, with poor pay and no prospects, and was hardly suitable for a man of thirty. He was surprised and not a little impressed when Forrest took it.

Forrest was always a keen shot and fisherman and he loved walking. Because he could not abide towns he walked six miles into Edinburgh each morning and back again at night. From nine to five he stood at a desk, scrutinizing and arranging dried and pressed plants sent in from all over the world. It would have been drudgery as tedious as a factory hand's job on a conveyor belt, if Forrest, with Scottish determination, had not determined to learn about and master the plants which at first he hated and which he came to love. Within two years he was the botanic garden's expert identifier and arranger of dried material; and then, by chance – that precious ingredient so necessary to 90 per cent of plant hunters – Sir Isaac was asked by a rich Cheshire industrialist to recommend a man who could undertake a botanical exploration in Western China. Cinderella's fairy godmother could not have done more for George Forrest. Sir Isaac highly valued his work in the herbarium but he would not keep him from such an opportunity. The Cheshire industrialist accepted the recommendation.

In 1904 Forrest left for Yunnan, the Chinese province 'South of the Cloud' which was to become his special territory. Augustine Henry had collected there before him, and other less-important plant hunters had collected sporadically and spasmodically in the area, but after 1904 North-West Yunnan was Forrest's preserve and he guarded it tenaciously. He was biting with those who considered that a plant hunter could flit like a moth from place to place and that no single collector had rights over an area; and he was thunderously menacing with poachers and claim-jumpers.

Right from the beginning he was enchanted by the province which was to become his home. North Yunnan wore a dramatic face; windswept downs, all at an altitude higher than any mountain in his native Scotland, between low ranges of mountain peaks and intersected by deep gorges. The rivers, which tumbled and stormed below, occasionally disappeared into subterranean courses. Cart traffic on the pot-holed tracks was possible, but only just. In the west Forrest found that the mountain ranges were so high and the rivers so deep that crossing either was a gargantuan task. In the south and east, where Augustine Henry had lived and where Forrest went to rest between seasons and between expeditions, he found green country where large lakes were common and fine crops were grown of Rice and Maize and Buckwheat and Opium Poppies*. He also liked the Yunnanese, who resembled the Scots in that they were hill people and wore homewoven cloth marked with special identifying patterns like highland tartans. Their mode of life, if a Western parallel could be made, was more French than anything; and this, again, was much to Forrest's taste. He never cared for

---

* Opium was being produced in large quantities when Forrest first went to Yunnan, though the anti-opium edict of 1906 brought the industry officially to a halt.

going into society, nor for entertaining: but when he was tricked into it he found Yunnanese society less formidable than similar gatherings in Scotland, and once he had become used to the order of dishes, he found the food more palatable than the tasteless and boring diet he had been brought up to. A meal in Yunnan began Scandinavian fashion with salty fish bits, moved on to rice with vegetables, especially the ubiquitous Radish and Turnip, then to eggs, duck or pork, and finished with a clear and delicately-flavoured broth of chicken and Mushroom which cleaned the palate. In his twenty-eight years there Forrest learnt to appreciate this classic order and, in the end, far preferred it to what he called 'dropping hunks of steak into a pond of brown Windsor'.

Everything, then, was to be near-perfect for George Forrest in Yunnan except that, when first he arrived there, the political situation was boiling up into an eruption. The unpoliced and uncontrollable marches had never been in a settled state of order, and the ordinary hostilities had been exacerbated by a British expedition to Lhasa under the command of Francis Younghusband, former political agent of Tonk. His idea was to negotiate a settlement with China and to settle relations between India and Tibet. In neither was he successful, and he made the great mistake of forcing his way into the Tibetans' holy city at the head of troops. They resented this so bitterly that after his withdrawal every white man in the frontierlands was automatically classed as a 'White Devil'. It was at this unpropitious moment that George Forrest arrived in Yunnan. But with Scottish obstinacy, British sang-froid and the average plant hunter's indifference to outside concerns, he refused to be worried by the threat of Tibetan guerrillas (who were generally Buddhist monks) and he set about the business of collecting.

There was no holding Forrest in such a delectable paradise for plant hunters. He established a permanent base at Teng-yueh and with a party of seventeen servants and collectors he set off north, moving up the river valleys of the Shweli and Salween then across the divide to the river Mekong. Eventually he reached the French Mission station at Tseku where in 1905 he set up a temporary headquarters as the guest of the old missionary-botanist Père Dubernard.

At Yaregong on the Tibetan border some distance away was one of the most remarkable of the French missionary-botanists, Jean André Soulié, a skilled physician, much loved by the local people, fluent in all the frontier dialects, and an inveterate collector who sent more than 7,000 dried specimens to Paris from the Tibetan marches. He was assisted there by yet another missionary-botanist, Père Bourdonnec. In the spring of 1905 they were warned by friendly lamas that a frontier dispute was developing between the Tibetans and Chinese in which 'White Devils' would stand very little chance of survival. The lamas had a long-standing affection for Soulié and begged him to leave instantly, but his year's botanical collections had not yet been sorted and packed for dispatch and Soulié temporized.

A few weeks later Père Bourdonnec crawled into Tseku. He was half-starved, half-naked, in a terrible condition. He told Père Dubernard and Forrest of the warning they had been given at Yaregong, and that, true to prophecy, the Mission had been attacked by Tibetan monks. He himself had managed to escape but Père Soulié had been taken prisoner. Dubernard had lived in the province many years and he found it difficult to believe they stood in any danger at Tseku, but then more news came from Yaregong. Poor Soulié had been tortured with increasing cruelty for fifteen long days and nights and then, mercifully, shot. Moreover, the same Tibetan monks who had murdered him were about to descend upon Tseku.

Père Dubernard ordered an immediate evacuation. Forrest's party added to the Mission staff made the total up to eighty. They fled. Twenty-four hours later the menace which at Tseku was so chilling seemed, somehow, to lose part of its reality. They caught an easy air of confidence from one another. Their pace slackened. There was even talk of turning back. It was at this point, bunched together in a river valley, that the Tibetan monks caught up with them.

Sixty-eight out of eighty were killed, amongst them all of Forrest's personal following save one. Soulié's assistant, Père Bourdonnec, having escaped from one massacre, was now shot through with poisonous arrows and while shrieking and writhing in agony was chopped to bits by double-handed swords. Old Père Dubernard was seized and tortured, quite literally, to death.

Forrest, new to the country, managed to escape. He hid up by day and tried to move south at night-time. The Tibetan monks hired women scouts to track the few who had escaped the massacre, and on the second day of his run Forrest discarded his boots because they made distinctive tracks. In forty-eight hours all he had to eat was a handful of parched peas and a few ears of wheat. The female scouts came so close to tracking him that he took to an icy stream, wading down its course with the water up to his chest. Despite this he was spotted and had occasion to thank God that he was little, for his hat was shot through with two poisoned arrows. By making a violent effort he outdistanced his pursuers, but they kept hard at his heels. He wrote later:

> At the end of eight days I had ceased to care whether I lived or died; my feet were swollen out of all shape, my hands and face torn with thorns, and my whole person caked with mire.

Starvation drove him to risk approaching a village. There, happily, the people were friendly towards him; but there, unhappily, they gave him parched barley to stay his hunger and this swelled up inside him causing him agony for hours.

Clutching at his belly he ran on, through cane brakes and along river torrents. Then it began to rain. The Tibetans were still on his trail. He saw them, and their torches at night-time. Slightly changing his direction he made a way for six days over glaciers, snow and ice and tip-tilted, jagged limestone strata which ribboned his feet. At last he reached what he thought must be a safe region, and there, in a path on the edge of a maize field, he stepped on a farmer's *panji* – a bamboo stake, sharpened and firehardened, stuck into the soil as a booby-trap:

> The spike, fully an inch in breadth, passed straight between the bones of my foot, and protruded a couple of inches from the upper surface.

He tore his foot free, stumbled on, and eventually found refuge in a Mission house in a town where Chinese troops were garrisoned. Another famous French missionary, Père Monbeig,* was also there, having escaped from his station in the north.

It took many months before Forrest's pierced foot was properly cured, and quite as long for him to trust the sight of a Tibetan. But from this harrowing honeymoon his relationship with Yunnan ripened into a fine and highly romantic marriage.

His success as a plant hunter and as a specialist in Rhododen-

* Nine years later Père Monbeig was chopped to bits by frenzied Tibetan monks in the station where Soulié had been murdered.

SOUTH OF THE CLOUD

drons was quite phenomenal. Indeed there were complaints that he was too efficient, too thorough, and sent home far too many seeds. Gardeners, however keen they were to raise new species, simply could not cope with the influx of seeds from Yunnan, especially as Kingdon-Ward and Joseph Rock were also sending huge quantities to Europe and America. Forrest's superb ability to delegate and organize was responsible for the size of his collections. Whereas most modern plant hunters preferred to take seed and plants for drying with their own hands, he adopted and developed Augustine Henry's idea of training and sending out collectors. He did it on a massive scale – as the executive, the chief-of-staff who received reports of discoveries, sent scouts to look over likely areas, and seed-gatherers at harvest time. His most enviable talent was for training his men: choosing them all from the same village so that they were related to one another and had a common loyalty to him; instructing them in what to look out for, how to cover the ground systematically, how to take specimens for drying, to mem- orize precise and detailed information about plant habitats. When they brought in seed at harvest time he superintended its drying, and he selected the best for sending to England. He also wrote up all the field notes in his own hand. As he was a trained botanist rather than a practical gardener some of his introductions were of small use to horticulture, but there was no denying his conscien- tiousness nor his ability to organize men and hold their loyalty. And it would be a mistake to think of him simply as a tented staff officer sending waves of private soldiers into the front line without doing any of the collecting himself. Far from it. He covered huge areas of ground, noting plants and their sites so that gatherers could be sent out at seeding-time. The machine he created worked so smoothly that it went on operating when he was on leave in England, and even after his death it took some time to run down. His friend at Teng-yueh, Rolla Rouse continued sending home packets of seed as they were brought in by the trained collectors.

Forrest's long and happy career in Yunnan could have met with no more satisfactory end than it did. His men had been collecting over a very large area and the season had been good. He wrote in a letter home from Teng-yueh towards the end of 1931:

When all are dealt with and packed I expect to have nearly, if not more than two mule-loads of good clean seed, representing some 4–5,000 species, and a mule load means 130–150 lbs. That is something like 300 lbs. of seed . . . If all goes well I shall have made a rather glorious and satisfactory finish to all my past years of labour.

While the last parcels were being packed he went snipe shooting. Because he had had pains in his chest he was carried out to the paddy fields in a chair and he sat on a low wall between the drowned fields while his bearers walked some distance away to flush the birds towards him. A noise made him look up. A snipe was drumming overhead, toppling at an immense speed and with that extraordinary throbbing sound of vibrating wing tips. Forrest leapt to his feet and fired. The snipe fell, and so did he; and both when they were picked up, were dead.

APPENDICES

Appendix I
Plant Distribution

lant frontiers are not easy to account for, nor are they easy to map. The world's divisions into climatic and physical regions, and the artificial boundaries imposed by difference of race, language, modes of living and thinking, hardly correspond at all to the frontiers of plant distribution. North America, for example, has been divided into four floristic regions: Arctic, Atlantic North American (divided into Northern and Southern), Pacific North American, and Caribbean. Yet climatically it includes twice that number of provinces. And no system of geo-botany can successfully account for the eccentricities of plant distribution when one Alpine meadow needs sometimes to be considered in separation from its neighbour or one section of a Burmese rain forest regarded as different from the rest, because they alone are endowed with unique botanical riches. There is the noted example of a line of division between the Arctic flora of America and the Arctic flora of Greenland. And, on a much smaller scale, there are tiny regions with unique floras which command the attention of serious plantsmen. There is the Great Orme for instance; a hill, no more, attached as if by accident to the Welsh seaside town of Llandudno which has a fascinating flora; and the small mountainous area of Gargano on the Adriatic coast of Southern Italy which is a positive Cinderella's Ball for botanists. Most fantastic of all is the rocky plateau above the Kaetuerk Falls in Guyana described by Mr A. Hyatt Verrill in his *Wonder Plants and Plant Wonders,* where he claims many strange plants are forced to a gigantic size by the warm, moist conditions. Amongst them are stout, Lily-like plants which collect water at the base of their eight-foot leaves, and in these vegetable pools live 'golden frogs and tiny silvery fish which have never been found anywhere else on earth'.

Three other points have to be borne in mind.

First, there is a curious and unaccountable similarity in floras between certain far-distant regions. Astoundingly, for example, there is a close connexion between the floras of the eastern sea-boards of the United States and those of eastern continental Asia and Japan.

Second, it is important to recall that floristic maps which inadequately express altitude are of little use. As all collectors have discovered on hills of any size, the vegetation and flora vary considerably at different heights up to the summit.

Third and last; however firm a plant frontier appears to be it must always be open to invasion. Seed dispersal by many ingenious methods from the explosion of released spring coils to burrs caught on clothing make all plant frontiers fluid, open to attack in force when a strong wind blows Thistle, Bullrush and Dandelion down, or by fifth column penetration through the alimentary canal of a single finch. If a similar habitat exists else-where and a seed can be naturally transported there without damage, the plant's original frontier is broken. It is on account of this that anomalies occur which make mongrels of most floristic regions and national floras. Plants from the Americas have been discovered in Ireland. Arctic plants have been found growing in the bland English West Country. Others, endemic to the Basque country, have popped up unexpectedly in Wexford, Cork and Clare. Wall Pennywort, a plant highly esteemed by the Japanese and truly native to the Orient, accommodates itself on the Mediterranean littoral and in the moist English West Country. White Bryony is a great favourite of English people, unknown to Americans, and hails originally from the tropics. In all these cases these frontiers have fallen and boundaries altered shape. Plant life has evolved.

Unhappily plant territories are sometimes artificially tampered with by speculators and exploiters. Inevitably there was a strong spirit of competitiveness when plant hunting was so profitable, and some very disagreeable things were done in the name of profit.

One particularly horrid illustration will do.

In 1870 when the Prussians showed they were determined to extend their influence in Europe far, far beyond the Pomeranian bog whence they had originated, one of their number, a professional plant hunter, bought from a tea-planter in high India a variety of the Slipper Orchid which grew in a small and limited area, and which – so far as he could see – was entirely unknown. It had a tinge of jade in its cream and violet blossoms, and was certainly worth buying. He took a number of specimens and nurtured them. Generously he named the Orchid *Cypripedium spicerianum*, after Mr Spicer, the tea-planter; then with unbelievable malignity this thorough son of Prussia ruthlessly destroyed the whole wild colony so that he alone would enjoy the distinction of introducing a unique Orchid into the markets of the West.

Inexcusable depredations such as this have never been common but they have added to the fluid state of all plant frontiers. Yet, in type, and bearing each of their weaknesses and limitations in mind, plant frontiers do exist, and it is this, and not the readily understood fact that grass is greener on the other side of every hedge, that has accounted for man's need to travel far and often under dangerous circumstances in pursuit of valued plants. When he has found them, noted a description, the exact location, the altitude, their associates, the nature of the soil and surroundings, he is faced with the problem of transporting his plants home, wherever that might be, and of recreating as far as possible the original habitat.

At its simplest an original habitat is recreated when an Apricot is grown in England sheltered by a south garden wall: and at its grandest when such a modern masterpiece as the Climatron in the Missouri Botanic Garden is made to house over 1,000 tropical plants in addition to a huge collection of Orchids. This is the most advanced of all attempts to create an artificial substitute for original habitats: a large geodesic dome made of light metal arranged in hexagonal patterns and, as well as plants, stuffed with mechanical apparatuses programmed by a Supervisory Data Center. Before a panel studded with buttons, switches and dials a supervisor sits – like God at the nerve-centre of creation – producing at will dry tropics, mist forest, tropical lowland jungle, and a bland oceanic climate; bounding, in fact, over a good many plant frontiers in one magnificent leap.

Appendix II
Plant Names

In its time plant-naming has aroused as much controversy as any other scientific subject.

Although Latin is internationally understood, a good many gardeners refuse to use botanical names under any circumstances. They regard it as both unnecessary and vexing to those who might otherwise be drawn in to share their pleasures. The more re-cherché amongst them would 'Englysshe' even the best-known botanic names such as Fuchsia, Dahlia, Hydrangea, Chrysanth-emum and Pyrethrum, though whether the 'Englysshed' or the Latinized version is easier to memorize is a matter of conjecture.

The disadvantages of using vernacular names make themselves especially obvious to anyone who talks or writes about a plant. English and American names for the same plant are quite often dissimilar. The English Fathen is Pigweed in the United States, Lesser Periwinkle turns into Myrtle, Wall Pepper is Love-entangle, the rare lace-like Orchid Creeping Ladies' Tresses makes the startling change to Dwarf Rattlesnake-Plantain. Then in different American states and different English counties 'local names' vary quite as much as local customs. Duckweed, for example, the ordinary *Lemna minor*, is less rich in equivalents than most English natives, but it still has ten, amongst them Boggart Creed, Jenny Green-teeth, Duck's Meat and Toad Spit.*

All these difficulties exist in the one vernacular language. They are multiplied many times for the amateur botanist who collects in foreign lands when changes of language, dialect and accent are

* Geoffrey Grigson, *The Englishman's Flora* (Phoenix House, 1958), gives the fullest list yet printed of English vernacular names.

frequent. That beautiful if noisome Iris, *I. foetidissima*, the Gladden or Roast Beef Plant, has many extraordinary European names: Xyris Puant in France, Giglio dei Morti and Flamma Fetida in Italy, Korallenschwertel in Germany, and its Turkish name has the ring of a bitter curse – Fena Kokulu Kuzgun Kilici.

There are, though, equal disadvantages in Latin. Possibly the most surprising is the exasperating fashion in which botanical names are altered with or without authority. Critics of the use of Latin have something to say when what they call Marsh Yellow Cress, named in Latin by Dr Candolle *Nasturtium palustre*, is renamed *Rorippa islandica* on the authority of two later botanists, Thellung and Schinz; when the Melancholy Thistle, called *Carduus heterophyllus* on the authority of no less a man than Linnaeus himself, has its name changed by Scopoli, an Italian, to *Cirsium helenionides*. And what did the great Reinchenbach think he was up to when he changed his own nomination *Habenaria chloroleuca* to *Platanthera chlorantha*?

Such alterations are inevitable as the science of botany matures and is more subject to accurate classifying; in other words, as it becomes altogether clearer to real botanists it becomes more and more opaque to the rest of us, and this is a pity because so many who might well enjoy collecting are too easily frightened away by the apparent labour of studying elementary botany and nomenclature. They may be reassured by a note on the purpose of using botanical names.

From the time of the early Greeks plants were named with a short description of their chief characteristics. The aim was to be brief, apposite and provide an easily-identifiable word-picture for other people to recognize. It was too clumsy a device to succeed, if only because descriptions of colours, scents, and even of form are largely subjective, but in particular because the economic use of words appeared to be beyond most botanists. Their names were tediously long and, as a means of identification, a spectacular failure.

Since then, chiefly because of Bauhin and Linnaeus, the vegetable kingdom has been divided and sub-divided again and again into an orderly pattern.

At its simplest, it breaks down like this:

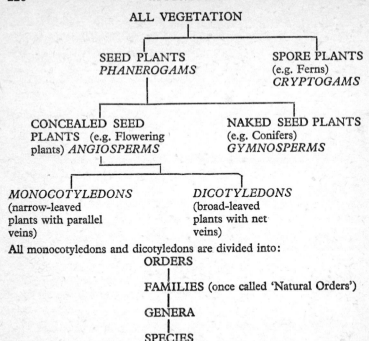

ALL VEGETATION

SEED PLANTS
PHANEROGAMS

SPORE PLANTS
(e.g. Ferns)
CRYPTOGAMS

CONCEALED SEED
PLANTS   (e.g. Flowering
plants) *ANGIOSPERMS*

NAKED SEED PLANTS
(e.g. Conifers)
GYMNOSPERMS

MONOCOTYLEDONS
(narrow-leaved
plants with parallel
veins)

DICOTYLEDONS
(broad-leaved
plants with net
veins)

All monocotyledons and dicotyledons are divided into:

ORDERS

FAMILIES (once called 'Natural Orders')

GENERA

SPECIES

Bauhin managed to impose on most of his contemporaries the excellent idea of economizing in words and giving each plant no more than two names, but it was Linnaeus who formalized the plan, and the two-word system was permanently established. The first, that of the genus, corresponds to our surname; the second, the equivalent of our Christian name, is used to describe the species within that genus. The comparison is more meaningful than whimsical. A human family has one surname, and each member his own distinguishing Christian name. So it is with a genus and, really, it is only necessary to know two names of any plant. The overfastidious would say this is insufficient. Species are as splittable as genera, and they add a third or even fourth name to show the distinction. These are the varietal names which should follow the abbreviation 'var.'; e.g., the Jew's Mallow with double flowers is *Kerria japonica* var. *flore-pleno*. Enthusiasts for this precise classification occasionally go to extraordinary lengths and are called 'splitters'; the rest of us, content with only two names, are called 'lumpers'.

Unlike plant names, which are generally printed in italics (and, nowadays, with no capitals for specific or varietal names), the abbreviation 'var.' is printed in ordinary Roman type. So is the abbreviation 'syn' for synonym (the renamers still cause sufficient chaos for it to be necessary to put alternative names) and the names of cultivars should also be enclosed in single inverted commas; e.g., from amongst the large number of Lily hybrids, *Lilium longiflorum* 'White Queen', and *L. philippinense formosanum* 'White Superior'.

Hybrids are sometimes natural. That is, they have occurred spontaneously through clandestine love matches rather than arranged marriages. The fruit of one such liason in 1797 was to produce an illegitimate but exceedingly beautiful Magnolia. In the garden of a M. Soulange-Bodin not far from Paris two species of Magnolia grew: a wine-violet *M. liliflora* and *M. conspicua* which has flowers of an uncertain white colour – almost that of dressed tripe. Between them they produced one seedling – the parent of all stocks of *M. soulangeana*.

But more often than not hybrids are the fruit of laboratory work, research and careful breeding by private enthusiasts or commercial nurserymen. Amazing garden flowers like the Sweet Peas of Mr Unwin and the Lupins of Mr Russell commemorate the hybrid-izers with their names. Other garden hybrids are named from their place of origin, e.g., *Cystisus kewensis*, a tom-thumb Broom which occurred unexpectedly at Kew, or from the chief area of their distribution, such as the French and African Marigolds of today. Both the last are, in fact, Mexican, but one was mainly introduced to Europe through North Africa and the other through France.

Many of the actual names used for cultivars show a strange taste for the commonplace or downright ugly. Whereas wild flowers and many garden plants enjoy beautiful and memorable names, some of the grandest achievements of the hybridizer's art have been saddled with the opposite. Can the French Marigold 'Miniature Lemon Drop' compare with Snapdragon, or the African Marigold 'Chrysanthemum-flowered Super Glitters' with Scarlet Pimpernel? An exquisite and unusual Lily deserves a better name than 'Corsage'; a Clematis with lavender flowers and slate-blue anthers suffers for ever as 'W. E. Gladstone'; three of the finest Geraniums ever produced are doomed as 'Fifth Avenue', 'Foxtrot' and 'Grand Slam'; and no one can begin to imagine what possessed the rosemen when they named their triumphs 'Café', 'United Nations', 'Rumba', 'My Fair Lady', 'Honeymoon' and 'Lilibet'. It seems unfair to flowers.

Linnaeus hardly ever made this mistake. He and his followers were 'lumpers': they had two primary objectives, to give a distinctive name to each genus, and an accurately descriptive specific name. With the former Linnaeus honoured the pantheon of Greece and Rome, the great men and heroes of the past, his friends and fellow botanists – though not all. With a salty sense of humour he named an American plant *Milleri* after his rival botanist Miller because the flower had a calyx which was 'close, short and completely enclosing the seeds'. The description suited Miller to a nicety. He too, was short, plump, and notoriously tight-fisted with information and with seeds.

With the specific name it was a question of describing the plant's outstanding characteristic, generally in Latin or in Latinized Greek. So *Stapelia bufonia* is named after the Dutchman, J. van Stapel, and it does look 'toad-like', and another species in the same genus is hairy, *S. hirsuta*.

A casual acquaintance with such words as *glaucescens* (bluish-green), *sylvestris* (of woods), *trichocarpa* (hairy-fruited), *officinale* (literally, 'of the shop', i.e., the apothecary's, and therefore, the type of herb he would naturally have in stock), *scutatus* (shield-like), *festalis* (gay), *ligulate* (tongue-shaped), etc; in fact any one of the more than 700* from *abbreviatus* (shortened) to *zeylanicus* (from Ceylon) does make botany as international as music and disease, and does add to a clearer understanding of the subject, and, in the long run, to the pleasures of collecting.

Botanical Latin is odd; that is, to anyone whose acquaintance with the language has been strictly orthodox. Yet, once it is appreciated, like Katisha's right elbow it has a fascination that few can resist. Whole books have been written about its history, grammar, syntax, terminology and vocabulary, and they contain gems of information. The Latin for certain place-names is marvellous: New Jersey turns out as Nova Caesarea: Pembroke as Transwallia; Ipswich is made declinable as Gippeswicum, and the Paiken Islands are honoured as Crocodilorum Insula. Much can be

* C. O. Booth has compiled a very useful list which appears on pp.453–459 of his *Encyclopaedia of Annual and Biennial Garden Plants* (Faber and Faber, 1957).

learnt about watery habitats from these books, that an ordinary marsh (*ulignosa*) is by no means the same as a peat bog (*turfosa*) or even a water-meadow which is periodically flooded (*inundata*), or a swamp which dries up like a sponge in summer (*paludes*). Then, if it gives pleasure to decline nouns with strange endings nothing could be more pleasurable than manipulating *bostryx* (*m.*). It has a musical ablative in the singular. By, with, or from a *bostryx* is simply *bostryche*.

Sometimes today it is actually necessary to decline botanical substantives; that is, on those dodo-like occasions when a new plant is found and named. When the discoverer or introducer has his name latinized to make a specific it is expressed either in the possessive-genitive case, e.g. *Clematis davidi* (David's Clematis), or as an adjective, e.g. *Prunus davidiana* (Davidian Peach). There is no strict rule about which use is correct, and following one or other practice has, in its time, caused trouble in botanical circles. No greater or more bizarre affair has ever occurred than in the struggle between supporters of *Victoria regia* and the supporters of *Victoria regina* in the naming of the Amazon Water Lily which took place over a hundred years ago. It involved eminent botanists of all nations, two learned societies, half the drawing-rooms of Potsdam, Paris and London, a collection of explorers, two dukes, and Queen Victoria herself.

Appendix III
Plant Collecting

Anyone with sufficient interest to go out and collect generally knows what he is doing. He will not gather a single specimen or one of a small colony if the species is rare. He is aware that plants are encouraged into more vigorous growth by being thinned, and – though a collector – he has the interests of the species much in mind and will never do a hundredth of the harm done to wild life by 'legitimate' despoilers such as farmers, roadmen and road-makers, and urban developers.

This is the answer to those cranky conservators who maintain that the only proper method of 'collecting' plants at the present time is with a camera. Besides, photographs of living plants cannot show their roots and other underground parts; a means of identification which is important in the case, for example, of some grasses. Nor, because lenses can play tricks and there are obvious difficulties in dimensions, can photographs be accepted as reliable. Nor will drawings do unless they are botanical sketches drawn with the aid of dissecting instruments and a microscope. A specimen of the plant has to be gathered, dried and pressed, and stored away as evidence of the find in a particular locality or for identification later by a specialist.

Collections can be arranged according to a variety of systems. The most usual is to classify them under each genus: or, if this is too large a number (there are over 500 genera in Britain alone), they may be arranged in families, (see p. 226). Physiological botanists will arrange their collections to show the functions of plant parts, and morphological botanists arrange theirs to show distinctive forms. And some will prefer to make collections of special types of plant: climbers, for example, or parasites or insectivorous plants. But some of the most interesting are made according to habitat; that is, arrangements of plants taken from both natural environments such as rock screes or river banks or heaths or wild woods or marsh saltings or sand-dunes, and from man-made habitats like cattle pastures, stone walls and roofs, railway sidings, staithes and sewage farms, motorway verges and wastetips. These divisions may be subdivided again into those which are wet and those which are dry, those which have acid soils and those which have alkali, and so forth.

EQUIPMENT IN THE FIELD

Botanizing can be an excessively costly business or one of the least extravagant of hobbies. The ways of illustrious plant hunters have made this clear. Linnaeus, for example, while collecting in Lapland was content with very little equipment and his own company. Joseph Hooker, on the other hand, took so much kit up into Sikkim and Assam that he needed a huge number of bearers to carry it from place to place, and, of course, he had to have a troop of sepoys to guard the lot.

The modern amateur botanist may be inclined to emulate Linnaeus or Hooker and be austere or luxurious in his fieldwork; but he will need to consider using some or all of the following:

i.   A collecting box – either a proper botanist's vasculum, japanned green or black, and with a snap lid like that of a snuff box; or simply a rigid, plastic container slung on a piece of string.

ii. If plants are to be pressed on the spot, either a portable press containing drying papers and tied with straps, or an old paper-back book held fast both lengthways and sideways by thick rubber bands.

iii. A supply of polythene bags for carrying aquatics and marsh plants, and specimens which are to be transplanted into the garden at home.
     It is also important to have lengths of thin, paper-covered wire to nip the necks shut.

iv. A trowel and pocket knife, the smallest possible pair of seca-teurs and a pair of scissors; or a type of bowie knife which can serve both as a cutter and digger.

v.   A watchmaker's lens for examining minute identifying marks.

vi. Some tiny numbered labels for attaching to plants. Notes on

each plant's habitat and precise location can be written against
the corresponding number in a small pocket book. A plant so
outstanding, so beautiful, or so scientifically intriguing that its
original whereabouts is instantly burnt on to the collector's
memory is uncommon; likewise photographic memories
amongst collectors are rare. And it is important to know a
plant's location if, later, seeds have to be gathered from it.

PREPARATION FOR DRYING

i.   *Cleaning*
      Dust off mud and insects and the soil from roots with a pastry
      brush.
      Remove dead, diseased, unshapely or surplus plants.
      If the plant is too large trim it down to size.

ii.   *Treatment of fleshy plants*
      Plants like Euphorbias, Brassicas and seaside species have to be
      treated like lobsters and thrown into boiling water. Otherwise
      they will continue growing in the press.

iii. *Aquatics*
      Limp water plants cannot be taken direct from a pool or they
      collapse and lose their form. Float them from the pool into a
      saucer. Then lift them out on a strip of thin paper to dry.

iv.   *Preserving colour*
      (a) Some plants have a tendency to go black when dried; e.g.
      Mints and Figworts. Iron them when fresh under a layer of
      brown paper with a hot flat iron.
      (b) The colour of orchids can most effectively be preserved by
      smoking them for ten minutes with a sulphur candle in a closed
      box. They look pallid when freshly smoked but the colours
      soon return and remain fixed for a long time.
      (c) With all other plants change the drying papers as often as
      possible. This helps, though it does not guarantee, to preserve
      colours.

DRYING AND PRESSING
Materials

i.   Drying paper: use commercial filter paper or unglazed tissue
      paper – but not, if possible, blotting paper or newspaper.

ii. Cotton buds and tweezers for arranging specimens; and Spencer-Wells forceps which click into a fixed position until released for dealing with brambles, etc.

iii. Thin pieces of glass to weigh down plants while arranging their parts.

iv. Ventilators – wooden slats or lengths of heavy-gauge wire – to place crosswise between layers of drying plants.

v. A press – either a botanical press with boards and straps, or an improvisation made of two plywood boards pressed down by scale weights or books or bricks.

vi. Thin cotton wool or lint for laying under thick flower heads so that they will be under equal pressure and remain unshrivelled.

Method

i. Cut off delicate flowers and dry them separately. They can be re-attached later to the stem.

ii. Drying takes from a few days to a fortnight.
Change the drying papers as frequently as possible. The first change should be made within twenty-four hours.

iii. As soon as papers are removed from the press, dry them out in an oven or in the sun.

iv. When all the specimens are dry, put them aside between store-sheets of newspaper.

MOUNTING AND STORING

Materials

i. Strips of music binding paper to stick down the plants, and a pure gum to attach small parts. It is wise not to use Sellotape, which is too strong, easily discolours, collects dust round the edges, and contracts drastically within a year or two.

ii. Reasonably heavy paper of a uniform size, quarto or folio.

iii. Snap-back folders or, better still, box files closed with tapes, for mounted specimens.

Method

i. Try to resist the temptation to make an attractive arrangement of mixed species on one page. It is easier in the end to mount one species only on a single sheet because it allows collections to be arranged systematically.

ii. Label each species as carefully as possible with its botanic and local names, habitat, exact location, and the date it was collected.

The information may be written on the sheet, or against a reference number in a separate notebook.

iii. Store the plants in a dustproof file or made-up cabinet and in a dry place.

Keep the collection free from insects with napthalene pads or mothballs.

A Select Bibliography

Not all the works referred to in the text, notes and appendices are listed in this bibliography. Those which do appear are of particular interest and are marked with an asterisk.

Allan, M.: *The Tradescants* (London, 1964)
Allan, M.: *The Hookers of Kew* (London, 1967)
Anderson, A. W.: *The Coming of the Flowers* (London, 1950)
Arber, A.: *Herbals, Their Origin and Evolution, a Chapter in the History of Botany, 1470–1670* (Cambridge, 1912)
Bean, W. J.: *Trees and Shrubs Hardy in the British Isles*, 2 vols. (London 1914)*
Bentham, G.: *Flora Hongkongensis* (London, 1861)*
Booth, C. O.: *Encyclopaedia of Annual and Biennial Garden Plants* (London, 1957)*
Botanical Mercury, The (Thomas Johnson) Vol. i (1634); Vol. ii (1641)*
Boyle, F.: *About Orchids* (London, 1893)*
Bretschneider, E.: *History of European Botanical Discoveries in China* (London, 1898)*

Carpenter, E. F.: *The Protestant Bishop; being the life of Henry Compton, 1632–1713, Bp. of London* (London, 1956)
Cox, E. H. M.: *Farrer's Last Journey* (London, 1926)
Cox, E. H. M.: *The Plant Introductions of Reginald Farrer* (London, 1930)*
Cox, E. H. M.: *Plant Hunting in China* (London, 1945)

Dampier, W.: *A New Voyage Round the World*, ed. A. Gray (London, 1927)*
Darwin, C.: *Journal of researches into the Natural History and Geology of the countries visited during the voyages of H.M.S. Beagle*, 2nd ed. (London, 1845)*
Darwin, C.: *On the Origin of Species* (London, 1859)*

Darwin, E.: *Botanic Garden* (London, Litchfield, 1789–91)*

David, A.: *Diary 1866–1869*, ed. and trans. H. M. Fox (Cambridge, Mass. 1949)

Dodge, B. S.: *Plants that have changed the World* (London, 1962)

Douglas, J.: *Journal kept by David Douglas during his travels in North America, 1823–1827* (London, 1914)

Elwes, H. J., and Henry, A.: *The Trees of Great Britain and Ireland*, 7 vols. (London, 1906–1913)*

Elwes, H. J.: *Monograph of the Genus Lilium* (London, 1880)*

Evelyn, J.: *Kalendarium Hortense*, 6th ed., 8° (London, 1676)*

Farrer, R.: *The Garden of Asia* (London, 1904)

Farrer, R.: *On the Eaves of the World* (London, 1917)*

Farrer, R.: *The English Rock Garden* (London, 1919)*

Farrer, R.: *Rainbow Bridge* (London, 1921)*

Fletcher, H. R.: *The Story of the Royal Horticultural Society, 1804–1968* (London, 1969)

Fortune, R.: *A Residence among the Chinese inland, on the coast, and at Sea* (London, 1857)

Fortune, R.: *Three Years Wandering in the North Provinces of China* (London, 1847)

Fortune, R.: *Yedo and Peking* (London, 1863)*

Gardener's Magazine, The (Ed. John Claudius Loudon) of 1833 and 1839

Gerard, J.: *The Herball*, F° (London, 1597)*

Good, R.: *The Geography of the Flowering Plants* (London, 1947)*

Goodspeed, T.: *Plant Hunters in the Andes* (London, 1961)

Gourlie, N.: *The Prince of Botanists, Carl Linnaeus* (London, 1953)

Gray, A.: *Observations upon the Relationship of the Japanese Flora to that of North America*, Mem. Amer. Acad. Arts and Sci: New Ser. (1859)

Green, J. R.: *A History of Botany in the United Kingdom* (London, 1914)

Grigson, G.: *The Englishman's Flora* (London, 1955)*

Hadfield, M.: *Pioneers in Gardening* (London, 1955)

Hagberg, K.: *Carl Linnaeus* (London, 1952)

Hagen, V. von: *South America Called Them* (London, 1949)

Harvey, A. G.: *Douglas of the Fir* (Cambridge, Mass., 1947)

Henry, A., and Elwes, H. J.: *The Trees of Great Britain and Ireland*, 7 vols. (London, 1906–1913)*

Herbst, J. F.: *New Green World, A Life of John Bartram* (London, 1954)

Hooker, J. D.: *The Botany of the Antarctic Voyage*, Pts. 1–3 (London, 1844–60)

Hooker, J. D.: *Himalayan Journals*, 2 vols. (London, 1854)★
Hooker, J. D.: *Outline of the Distribution of Arctic Plants*, Trans. Linn. Soc. Lon. (1861)

Kaempfer, E.: *History of Japan*, 3 vols. (London, 1906)★
Kew, H. W., and Powell, H. E.: *Thomas Johnson, M.D. of Selby* (London, 1932)
Kingdon-Ward, F.: *The Romance of Plant Hunting* (London, 1924)
Kingdon-Ward, F.: *Pilgrimage for Plants* (London, 1960)

Lemmon, K.: *The Golden Age of Plant Hunters* (London, 1968)
Linnaeus, C.: *Flora Japonica* 8° (Amsklaedami, 1737)★
Linnaeus, C.: *Species Plantarum* (Holmiae, 1753)★
Livingstone, J.: *Observations on the Difficulties which have existed in the Transference of Plants from China*, Hort. Trans., Vol. III (1820)★
Lloyd, C.: *William Dampier* (London, 1966)

McClintock, D.: *Companion to Flowers* (London, 1966)
Millican, A.: *The Travels and Adventures of an Orchid Hunter* (London, 1891)★

Ohwi, J.: *Flora of Japan* (Washington, D.C., 1965)
Oliver, F. W.: *Makers of British Botany* (Cambridge, 1913)
Oliver, S. P.: *The Life of Philibert Commerson*, Ed. by G. F. Scott Elliott (London, 1909)★

Parkinson, J.: *Paradisi in sole Paradisus terrestris* (London, 1629)★
Parkinson, J.: *Theatre of Plants* (London, 1640)★
Porter, C. L.: *Taxonomy of Flowering Plants* (San Francisco, 1959)

Raven, C. E.: *John Ray, Naturalist, his life and works* (Cambridge, 1942)
Ray, J.: *Catalogus plantarum Angliae, et insularum adjacentium* 8° (Londini 1670)
Ray, J.: *Synopsis methodica stirpium Britannicarum*, 8° (Londini, 1960)
Robinson, W.: *The Wild Garden* (London, 1870)

Sargent, C. S. *Plantae Wilsonianae* (Cambridge, Mass., 1913–17)★
Siebold, P. H. von: *Manners and Customs of the Japanese* (London, 1841)
Sighart, J.: *Albert the Great, of the order of Friar-preachers, his life and scholastic labours.* From original documents by J. S., trans. from the French by T. A. Dixon (London, 1876)
Smith, E.: *The Life of Sir Joseph Banks* (London, 1912)
Spruce, R.: *Notes of a Botanist on the Amazon and the Andes* (London, 1908)

Standish and Noble: *Ornamental Plants and Planting, Practical Hints on* (London, 1852)
Stearn, W. T.: *Botanical Latin* (London, 1966)

Taylor, G.: *Some Nineteenth Century Gardeners* (London, 1951)
Tergit, G.: *Flowers through the Ages* (London, 1961)
Turner, W.: *A New Herball*, F° (London, 1551)*
Turrill, W. B.: *The Royal Botanic Gardens at Kew* (London, 1959)
Turrill, W. B.: *Joseph Dalton Hooker* (London, 1964)

Veitch, J. H.: *Hortus Veitchii* (London, 1906)
Verrill, A. H.: *Wonder Plants and Plant Wonders* (New York, 1939)*

Wallich, N.: *Upon the Preparation and Management of Plants during a voyage from India*, Hort. Trans., Vol I, 2nd Series (1835)
Ward, N. B.: *On the Growth of Plants in Closely Glazed Cases* (London, 1842)
Wilson, E. H.: *A Naturalist in China with Vasculum, Camera and Gun*, Intro. by C. S. Sargent (London, 1913)

Index

Plant names mentioned only in Appendix II are not indexed.

The Mountain People 75p
Colin Turnbull

'I might compare the book to *The Godfather*, if Turnbull's study were not more gruesome, more frightening in its implications for us all. Within the Mafia, after all, there is a morality. The Ik have none . . .'
Robert Ardrey

One of the most remarkable and acclaimed works of contemporary anthropology, this is a devastating portrait of the Ik, a black African tribe, in the final stages of decline and dissolution.

The Ik, a tribe of nomad hunters, were driven from their hunting grounds into the barren mountains that separate Uganda, the Sudan and Kenya. Forbidden to hunt the animals that once fed them, they were asked to become farmers in a land with no rain and with no knowledge of farming.

The results were disastrous. In less than fifty years all social conventions have gone and family life has totally disintegrated. Children steal food from the old, parents starve their children to death, the aged and frail are abandoned to die. Only the toughest and most cunning survive.

In this brilliant and penetrating study of the Ik, Colin Turnbull relates their experience to modern society: the conclusions he draws are as savage and disturbing as those of *Bury My Heart at Wounded Knee*.

'This book is one that no reader is likely to forget, however comfortably he may otherwise toast his toes before the fire of life'
Guardian